W9-CUZ-076

FUELING
THE
FUTURE

FUELING
THE

EDITED BY **ANDREW HEINTZMAN**
AND **EVAN SOLOMON**

FUTURE

HOW THE BATTLE OVER ENERGY
IS CHANGING EVERYTHING

ANANSI

Copyright © 2003 Realize Media

All rights reserved. No part of this publication may be
reproduced or transmitted in any form or by any means,
electronic or mechanical, including photocopying, recording,
or any information storage and retrieval system, without
permission in writing from the publisher.

Published in 2003 by
House of Anansi Press Inc.
110 Spadina Avenue, Suite 801
Toronto, ON, M5V 2K4
Tel. 416-363-4343
Fax 416-363-1017
www.anansi.ca

Distributed in Canada by
Publishers Group Canada
250A Carlton Street
Toronto, ON, M5A 2L1
Tel. 416-934-9900
Toll free order numbers:
Tel. 800-663-5714
Fax 800-565-3770

07 06 05 04 03   FP   1 2 3 4 5

National Library of Canada Cataloguing in Publication Data

Fueling the future : how the battle over energy is changing everything /
Andrew Heintzman and Evan Solomon, editors.

ISBN 0-88784-695-5

1. Fuel. 2. Power resources. I. Heintzman, Andrew II. Solomon, Evan, 1968–

TP318.F84 2003   662'.6   C2003-905895-6

Cover and text design: Bill Douglas at The Bang
Cover photos: (top) Bill Douglas, (bottom) Getty Images
Page composition: PageWave Graphics Inc.

Canada Council   Conseil des Arts
for the Arts   du Canada

*We acknowledge for their financial support of our publishing program
the Canada Council for the Arts, the Ontario Arts Council, and the Government of Canada
through the Book Publishing Industry Development Program (BPIDP).*

Printed and bound in Canada

For Molly, Theodore, and Maizie

CONTENTS

PREFACE

The virus spread through the city with alarming speed. By early May dozens were dead. Death brought panic, and panic brought paranoia, until the mere act of sneezing was considered dangerous. The streets emptied of their usual crowds and businesses promptly closed. Physicians were helpless, offering neither a cure for the virus nor an idea about its origin — though the source was probably China. The politicians were worse than helpless. Weakened by years of infighting, the leaders were so self-absorbed in their efforts to cling to power that all they could do was offer their usual tepid medicine: rhetoric. It was, to use Dickens's phrase, the worst of times. At home and across the globe, religious tensions threatened to break out into wider conflict — a clash of civilizations, as it were. It seemed as if all the conventional ideas about how to solve the problems facing the world were failing at once. And in the midst of this chaos, Henry Oldenburg, an affable German immigrant, put his faith in ingenuity and changed everything.

It was an astonishing act of optimism. After all, Oldenburg was a man of moderate qualities. He was not a great scientist, a brilliant physician, or a thinker of any note. Charming, fluent in several languages, and well travelled, Oldenburg began his career making an inconspicuous living as a tutor. But through one of his wealthy pupils, he managed to get himself appointed as the secretary of the newly formed Royal Society of scientists. He watched some of the world's great thinkers sharing ideas, testing inventions, and making discoveries. And then Oldenburg had an epiphany. What if he could collect all the ideas of these great minds and share them regularly with the public? Would that be the first step towards, as he would later put it, the "Grand Design of improving" the world? He believed it would. And so, in the very months when the ghastly virus was carving its way through the population of London, a newsletter was quietly being distributed. The mandate, as Oldenburg wrote on the opening page, was to gather "solid and useful knowledge" about "ingenious Endeavors." The year was 1665. Henry Oldenburg had just published the very first scientific journal, called *Philosophical Transactions*.

It should be said that the sixteen-page periodical did not immediately have a seismic effect on the world. Certainly not when the bodies of 15 percent of London's citizens were literally piled in the streets, their tongues swelled, their skin black and covered with horrible sores, or buboes, from the bubonic plague. Nor did the newsletter have a notable effect on the newly restored king, Charles II. Only five years had passed since the demise of Oliver Cromwell's radical democratic experiment called the Protectorate, and dandy young Charles was too busy waging a low-level war with Anglican parliamentarians to study big ideas. But then, who today has time for such things? Even back in the seventeenth century, news events were moving so quickly that stepping back to gain new perspective seemed like a luxury no one could afford. If you survived the Black Death of 1665, as Henry Oldenburg did, you would hardly have a moment to relax before the Great Fire of 1666 destroyed two-thirds of London. In the wake of these events, an apocalyptic fever seized the country.

Clergy announced to a thoroughly petrified public that the end was nigh. A radical Quaker named Solomon Eccles ran naked through Cork Cathedral screaming about the coming doomsday. Irrationality, not ingenuity, was the order of the day. Oldenburg's little newsletter was clearly not the solution the public craved.

Yet he persisted. Year after year, his journal chronicled in growing detail the solid and useful knowledge about the ingenious endeavours of the Royal Society, and slowly this knowledge began to transform the world. Anton van Leeuwenhoek's invention called the microscope, for example, first described in *Phil. Trans.*, soon revolutionized science and medicine. Isaac Newton's early experiments on light with the prism and his subsequent invention of the calculus (a claim that caused a delicious scandal when Gottfried Leibniz declared that he pioneered it first) took mathematics, engineering, and astronomy to a new level. Since then, a host of famous scientists, from Alfred Whitehead to Stephen Hawking, have announced their work in *Phil. Trans.*, making Oldenburg's little periodical the oldest and arguably the most influential living magazine in the world.

The fact is, Oldenburg would probably feel very much at home here in modern-day Canada. There's no bubonic plague, thankfully, but deadly viruses still afflict us and elude our treatments. As we write this preface, the funeral for Nelia Laroza, the first health-care worker to die of SARS, is taking place in Toronto. Meanwhile, the discovery of a cow with BSE — mad cow disease — has punctured the agriculture economy of Alberta, and the rest of us are bracing for another outbreak of West Nile virus. The political landscape is also reminiscent of Oldenburg's time. The Christian religious wars that he knew have been replaced by the so-called "war against terror," which politicians are too careful to admit is actually a war against Islamic fundamentalism or Islamic fascism. The terrors in London during Cromwell's time have given way to terror of a different sort, suicide bombings everywhere from Moscow and Jerusalem to Bali. Governments too are acting with remarkably similar authority to those in the seventeenth century — as Oldenburg himself might mention. After all, for a brief time he was actually imprisoned in the Tower of London for the mere act of writing letters to a French scientist. It was an absurd abuse of power by a regime not yet versed in the modern theory of basic human rights, but then again, over 600 people are currently being held without trial or legal representation in Guantanamo Bay, Cuba, by the American military. So 2003 is really not so far from 1665 after all.

The best ideas often emerge over a drink, which is why about a year ago we asked some friends and colleagues to meet at a bar. They were a diverse group: one worked for Médecins Sans Frontières, another for a large media conglomerate. We were joined by a student, the CEO of an investment company, and a cell-phone executive based in Bucharest. Like Oldenburg, we were interested in how the ingenious endeavours of our time might solve the dilemmas facing the future. We asked everyone two simple questions: What are the problems that concern you the most? Are there solutions to fix them? That got the drink tab started and the ideas flowing.

One person was worried most about infectious diseases here and in the developing world. Another was interested in environmental issues and the effects of climate change. Another was concerned about poverty and yet another with the ineffectiveness of government. The list went on and on. But to all of our surprise, for each problem there was a possible solution: low-priced generic drugs were available for AIDS patients in developing countries; ideas like full cost accounting and new technologies like hybrid cars might help alleviate certain environmental problems; microbanks were providing another tool to fight poverty. Everyone had examples of someone supplying a solution to help solve one of these seemingly intractable problems. The mood shifted from one of despair to excitement, and we were quite sure the drinks played only a minor role in the change.

As the conversation in the bar got more intense, it became clear to everyone that the challenge was not necessarily to *invent* the solutions to the problems, rather it was to create a link for people who were already out there *solving* them. But what would that link be?

It was our friend Thomas "Tad" Homer-Dixon who had provided the answer, in his best-selling book *The Ingenuity Gap*. He believes that as societies grow more complex, a gap develops between the demand for ingenuity and its supply. The challenge is to close that ingenuity gap. That was our link. Is it possible to supply the right amount of ingenuity to the problems facing us in the future? We all concluded that it is.

The Ingenuity Project is an annual effort to bring together bright minds from around the world in order to provide practical and ingenious solutions to the complex problems facing the world. The subjects will be as wide-ranging as the problems confronting us: environmental issues (energy, climate change, water), economic issues (globalization, Third World debt), political issues (clashes of civilizations, AIDS, efficient government) and legal issues (global terrorism, world courts). This year the topic, and hence the title of this book, is *Fueling the Future: How the Battle Over Energy Is Changing Everything*.

Why start with energy? Before August 14, 2003, the answer to that question might have included a whole range of subjects. We might have talked about Middle East politics or the Kyoto Protocol, we might have cited statistics about the influence of the auto industry and the price of gas, or we might have presented facts about climate change and the health hazards of smog. After all, there is plenty of evidence to support the notion that energy is at the heart of the world's most critical problems and concerns.

But with the blackout of August 14, when over 50 million people across eastern North America lost power, the question of energy went from the theoretical to the urgent. For many of us, when the lights went out, normal life came screeching to a halt. People living and working in high-rises could not use elevators, so they had to either stay at home or walk up and down dozens of flights of stairs. At hospitals, generators kept critical-care facilities running but the research labs with their coolers and temperature-controlled environments had no backup, putting high-level research projects in jeopardy. Surgeries stopped. Gas stations ran out of fuel after first spiking the price. Businesses closed. Parliament Hill went virtually dark. Cell-phone service broke down. If the blackout had lasted for more than forty-eight hours, Toronto would have run out of water. With no lights, no traffic control, no clean water, no food on store shelves, no electricity to cook or wash clothes, no gas to drive cars or pick up garbage, no businesses running, and the government virtually paralyzed, all the generosity we saw in the first few hours would have given way to Darwinian survival instincts. In other words, even a simple malfunction in our complicated energy grid could turn our largest cities into places not much better off than Baghdad. Our way of living is based on something that fragile.

It's a lesson we tend to forget. Thirty-eight years ago, another blackout raised a similar alarm. On November 9, 1965, at 5:27 p.m., 30 million people across the Northeast, from New York to Toronto, lost power for fourteen hours. The cause back then was also due to a "cascading effect": a single failure at the Sir Adam Beck power station in Ontario triggered a rolling shutdown across the grid. In 1965 General Electric used to publicize the slogan "Live Better Electrically." With the lights out, people suddenly realized how true that slogan was. Energy means life. Period.

So it's no surprise that after the summer blackout of '03, obscure terms like megawatts, power grids, and peak usage became part of North Americans' popular vocabulary. But a reliable electricity grid is only part of the energy question. There are many more questions about energy — questions about climate, health, economic prosperity, and sustainability. Not surprisingly, given that an election in Ontario was a matter of months away, the politicians and pundits rushed into the fray and played the blame game. Some said the blackout was caused by privatizing the electrical grid, others claimed it was caused by Canada's dependence on American energy, while still others said it was caused by our lack of energy supply. But the bigger questions were lost in the fury. Before investing billions of dollars into quick-fix, politically expedient solutions, we should step back and ask what kind of energy system as a whole we ought to have. What is the best way to produce cheap, reliable energy that will be sustainable for our future and for the planet? In other words, what is the best way to fuel the future?

In the midst of the political sloganeering, the panic of the blackout, the smog days, and the oil wars, we've gathered some of the best minds and sharpest stakeholders in the energy field to answer these broader questions. Following the lead of Henry Oldenburg, we've asked them to supply ingenuity about the future of energy. And their answers are extraordinary. It turns out that the energy world is in a dramatic transition period, and harnessing that ingenuity may well allow us to solve the energy crisis in new and imaginative ways.

But there's a catch. As Oldenburg first realized when he published *Philosophical Transactions* in 1665, scientists and big thinkers often speak a language that very few of us can understand. In his day, scientific works were written in Latin, and so Oldenburg insisted that all his contributors write instead in the language of their land. So in that spirit, we've made sure that all our contributors write in plain language, not in academic or scientific jargon. After all, everyone has a stake in the future, and so everyone — not just the experts — ought to be able to make the right choices about the route we're going to take and the solutions we're going to use.

And innovative solutions do exist, and are being developed and implemented at this very moment. For every problem, there's a new generation of thinkers, inventors, visionaries, and engineers who are already beginning to supply the kind of ingenuity needed to solve it. Most of us, however, don't even know what these ideas are, let alone what they really mean or how they are being applied. That's why we see this as a learning process. It's about asking the basic questions about complicated problems and getting the answers needed to take action. So the first job is to provide us all with a clear menu of better options.

Still, how do you do that without becoming overwhelmed or polarized into endless debate between two opposing sides? To avoid this trap we've asked all the participants to tackle the issues from a governing perspective, not a critical or a theoretical one. As governors we are all forced to build consensus, to be practical, to consider the consequences of our choices both for the current and the next generation. In other words, as governors we can build an action plan for the planet. And that action begins at the most fundamental level, with the energy we use to survive.

INTRODUCTION

On Thursday, June 12, 2003, fifteen kilometres from the town of Baiji, an enormous fireball rose into the sky like a giant balloon and exploded. Scorching oil drops rained down on the residents as they fled for cover. "We heard two explosions and ran," Abu Ala, the owner of a nearby cafe, told an AFP correspondent. "Then we saw fire shooting out of the pipeline in two places."[1] Minutes later Abu Ala heard another sound, the deep hack of two American AH-64 Apache helicopters circling overhead, their 33-mm M230 guns swivelling like bloodhound noses trying to pick up a scent. There was nothing to find besides the carnage, but the choppers' presence made sure the area was secure, at least for the moment.

This was supposed to be a better day in Iraq. Only hours earlier the U.S.-led Iraqi administration awarded the first post-war contracts to an array of multinational oil companies. The crude that was spraying poisonous flame into the air was meant to be heading safely to markets like Turkey. Flowing oil is the international language of progress, and the Americans wanted to send the world a signal: money will come in, Iraqi cities will be rebuilt, and life will get better. The United States wanted to erase memories of petroleum fires and the black rain that accompanied them. Those were from the first Gulf War. This time things were supposed

1

to be different. The president of the United States promised.

Unfortunately, the saboteurs weren't listening to President Bush. They chose their target as carefully as they chose the date of their attack. Baiji is a major refining and processing town, using the oil from the fields around the northern city of Kirkuk to generate electricity for central and northern Iraq and, more importantly, Baghdad. The assault set back America's effort to rebuild the country — an effort costing over US$1 billion every seven days — by weeks.

For all this, the attack didn't come as a surprise. The American military was well aware that the pipeline was a target. Baiji lies in the centre of Sunni territory, the Sunni are the Muslim group that once dominated Saddam Hussein's Bathist regime, and Saddam still had strong support in this part of the country. All of which explains why the U.S. military also carefully chose this day to launch what Lieutenant General David McKiernan called "a very decisive, very lethal" assault on "terrorist training camps" in the north. Despite the loss of an Apache helicopter in that attack, McKiernan called the operation a success. But as the sky around Baiji blackened, everyone knew that if the war against Saddam's regime was over, the battle over his oil had just begun.

Ten days later another explosion outside of the town of Hit announced the next target. Hissing like a giant snake, an orange flare from one of the biggest natural gas pipelines in the country slithered into the sky. The fire was so large that it burned for a day before anyone even tried to control it. The pipeline was a key supplier to the electricity generation plants for Baghdad. Knocking the line out was crushing for the already beleaguered capital city. "People are already living in hell," Dathar al-Qassab, the head of the refinery that supplies the pipeline, told the *Globe and Mail*, "and it's only going to get worse."[2] Once again, the attack was carefully timed to do both economic and symbolic damage. At the very moment the bomb went off, U.S. officials were celebrating the first Iraqi oil to be sold abroad since the beginning of the war three months earlier. In the Turkish port of Ceyhan, a tanker named *Ottoman Dignity* was pumped full with a million barrels of crude and set to sea. This was supposed to be yet another American signal to the Iraqi people that oil was flowing once more and order was restored, but again, the signal was scrambled. In any event, the ceremony was more of a PR stunt than anything else. Even those people at the ceremony knew there was something of a Potemkin village aspect to what they were seeing. That's because all the oil inside the *Ottoman Dignity* actually came from a storage facility that was filled long before the war started. The pipelines meant to deliver *new* oil to Ceyhan were destroyed in the attack ten days earlier. In other words, very little oil was flowing out of Iraq at all.[3]

That oil is at the centre of global conflict is not new. Since that Saturday afternoon on August 27, 1859, when Edwin Drake struck oil in Titusville, Pennsylvania, the heavy black liquid has motivated wars, built empires, and transformed society. Just think of it: immediately after Drake's discovery the Age of Illumination arrived, as the first oil was refined into kerosene and used to light homes. Near the turn of the century the age of the automobile arrived, and Henry Ford's Model T and the internal combustion engine soon altered factories and mobilized the once stationary public. Cities expanded into suburbs, which hemorrhaged into urban sprawl. Emerging oil barons like John D. Rockefeller of Standard Oil and Marcus Samuel and Henri Deterding of Royal Dutch/Shell gained wealth that rivalled nations'. By the eve of World War I, when the first lord of the admiralty, a young Winston Churchill, converted Britain's Royal Navy from coal to oil, the link between power and oil was solidified.

World War II, however, illustrated the proximity of that link in much more dramatic fashion. On June 22, 1941, Hitler ignored all military advice and attacked his former ally, the Soviet Union, opening up a second front. With three million men, 600,000 motorized vehicles, and more than half a million horses, Operation Barbarossa had many objectives, but chief among them was the capture of the oil fields of the Caucasus, especially those around Baku. In his masterful book *The Prize: The Epic Quest for Oil, Money and Power*, Daniel Yergin explains the logic behind the Nazi's plan. Hitler had always realized that energy self-sufficiency was the key to his empire's survival, but Germany had limited access to oil reserves. Hitler ordered the chemical giant IG Farben to develop a massive supply of synthetic fuel, but even employing hundreds of thousands of slave labourers at Monowitz — the huge industrial complex located inside the Auschwitz concentration camp — IG Farben could not meet the demand. The attack on Russia was inevitable. "Unless we get the Baku oil," Hitler told Field Marshal Erich von Manstein in 1942, "the war is lost."[4] He was right. Russia held on to its oil fields at a terrible cost, and Hitler's prized divisions, both the Sixth Army at Stalingrad and Rommel's Afrika Korps in North Africa, literally ran out of gas and surrendered. It was the beginning of the end of the Third Reich.

Germany's ally Japan was in a similar situation. Like Hitler, Prime Minster Hideki Tojo realized that lack of natural resources was the chief weakness of his small island nation. His solution lay in capturing the rich oil fields in the East Indies. It was a dangerous proposition. Conquering the Pacific Rim and invading the East Indies would likely trigger a military response from America, whose powerful Pacific fleet at Pearl Harbor was a threat to the Japanese flank. Tojo gambled, however, that a decisive blow to America's navy would knock the United States out of the region permanently, or at least long enough for Japan to secure its supply of oil. He began planning an audacious attack.

Ironically, right up until the outbreak of war, Japan

imported most of its oil from, of all places, the United States. That trade relationship was deteriorating quickly, however, as Japan's ongoing conflict with China and expansionist policy along the Eastern Pacific Rim grew more aggressive. Western countries were horrified by reports of Japanese atrocities in places like Nanjing and Chongqing but were too busy dealing with Hitler in Europe to do more than slap trade embargoes on Japan. Fearing that strict sanctions on oil might spur it to widen its war, President Franklin Roosevelt kept crude flowing to Japan until the very last moment. It was an error in judgment.

On December 7, 1941, just before eight o'clock in the morning, the first of Admiral Isoroku Yamamoto's 353 planes dropped their bombs on Pearl Harbor. Though the military gamble initially paid off handsomely for the Japanese, it backfired in its grander strategic assumptions. America did not, as Tojo hoped, limp away from the Pacific region and leave its resource treasures to the Japanese. Rather, with its massive manufacturing base and access to huge supplies of domestic oil, the United States regrouped, rebuilt, and attacked. The Japanese quest for oil lost them an empire and led them to disaster.

It's a pattern that has been repeated throughout the last century. In August of 1990, Saddam Hussein made the very same error as Tojo. With a massive foreign debt accumulated from his calamitous war against Iran, Hussein needed to increase his oil exports keep his government afloat. The conquest of oil-rich Kuwait, he believed, was the answer. Saddam's strategy was familiar: a quick and decisive victory in Kuwait would undermine any military response from the weak-kneed West, which would be forced to negotiate a new and, for Hussein, favourable peace. He was wrong.

Seventeen years earlier another war in the region taught the West a hard lesson about how integral Middle East oil is to the health of the global economy. It was October 6, 1973. While

Israelis were celebrating Yom Kippur, the holiest day of the Jewish year, Egyptian tank divisions along the Suez Canal and Syrian tank divisions in the Golan Heights rolled across Israel's southern and northern borders. Taken by surprise, Israel fared badly in the initial stages of the war, but it quickly reversed the circumstances. Eventually both the Golan Heights and the Sinai Desert fell under Israeli control and the Arab countries were forced to sue for peace. It was a humiliating loss for the Arab world.

To punish the West — specifically America — for its support of Israel during the war, the thirteen-year-old Organization of Petroleum Exporting Countries (OPEC) hiked the price of a barrel of oil by 70 percent and put an oil embargo on the United States. Only eleven days had passed since the war in the Middle East broke out, and with shocking speed its effect reached the shores of North America. The first modern energy crisis had arrived. The United States was much more vulnerable than it realized. Energy self-sufficient twenty years earlier, the country now imported around 35 percent of its oil. The embargo sent the stock market spiralling down and inflation rates rocketing up. Within days there were long lineups at gas stations across North America. By November President Nixon banned buying gas on Sundays and asked stations to pump a maximum of ten gallons (38 litres) per car. Environmental laws were loosened so that alternatives like coal and nuclear could supplement the supply of energy. All across North America the relationship between energy, the economy, and everyday life was suddenly seen in high relief. The consequences of this realization were enormous. It kick-started the environmental movement, the alternative energy industry, and massive research and development into more efficient existing technologies. But more than anything else, it illuminated the central role that oil played in keeping the global economy robust.

Which is why in 1991 George Bush Sr. did not have any difficulty convincing the leaders of thirty-nine other countries to join what he called Operation Desert Storm. On February 24, 670,000 troops marched into Kuwait, and within 100 hours Saddam was in retreat back to Baghdad. As we all know, the story doesn't end there. On March 19, 2003, the United States and a rather more anemic coalition of allies once again went to war against Saddam. This time their mission priority was obvious: capture and secure the main Iraqi oil fields before Saddam can destroy them. Three months after the war ended, with Saddam still eluding capture and with no weapons of mass destruction found, gaining control of the oil fields is the only part of the coalition's mission deemed a success, and as we've seen, a precarious one at that.

The story of oil, our most important energy source, is also an economic story, as the business of oil has fundamentally changed our modern economies. To see the evidence of this, you only have to look at *Forbes* magazine's ranking of the largest companies in the world. Three of the top ten companies in the world in 2002 — ExxonMobil, Royal Dutch/Shell Group, and BP — are energy companies. Their combined revenues are in excess of $600 billion. But none of these three companies, no matter how large and powerful, can match the incredible strength of the original energy behemoth, Standard Oil.

Perhaps no other company has had such a far-reaching impact on the twentieth century as Standard Oil. From the time of its founding in 1868 until 1911 when it was found to be a monopoly and broken up, Standard came to define the exploding power and wealth of the new oil age. The company was founded, driven, and animated by arch-capitalist John D. Rockefeller. Shortly after the strike in Titusville, Rockefeller saw the commercial potential of oil. But he chose to invest in the refineries, a strategy he believed was less risky than investing directly in oil exploration. With his base in Cleveland, Ohio, Rockefeller and

Standard were poised to take over the burgeoning oil industry, and by 1885 the company controlled 90 percent of all of the oil refined in America. Rockefeller aggressively used this monopoly position to drive his competitors out of business and pick up their pieces when they failed.

At the time it was disbanded, Standard was — in the parlance of today — a vertically integrated trust that largely controlled the refining, production, marketing, and transportation of oil in the United States. The antitrust trial of 1911 helped to define anticompetitive commercial practices and was a precursor to later trials of AT&T, IBM, and Microsoft. Even after Standard Oil was broken up, its spirit was to reverberate throughout the twentieth century, with twenty-four companies continuing to carry the Standard name. The descendants of the Standard empire still include many of the most powerful companies in the world, including ExxonMobil and Imperial Oil.

But the influence of petroleum on our lives is far more intimate than suggested by economic and geopolitical stories. It is virtually impossible to imagine everyday life without it. The American Petroleum Institute likes to publish an alphabet of products that depend on petroleum, and it's worth taking a look at what sort of products are included before we try to engage in the difficult task of imagining a different kind of energy future. Petroleum is crucial to (among other things) antihistamines, antiseptics, audiocassettes, baby strollers, balloons, bandages, cameras, candles, CD players, computers, credit cards, dentures, deodorant, diapers, digital clocks, dyes, eyeglass frames, fertilizers, food preservatives, food storage bags, furniture, garbage bags, glue, golf balls, hair dryers, heart-valve replacements, house paint, infant seats, ink, in-line skates, insecticides, life jackets, lipstick, luggage, medical equipment, nylon rope, pantyhose, perfumes, photographic film, piano keys, roofing, shampoo, shaving cream, soft contact lenses, sunglasses, surfboards, surgical equipment, syringes, telephones, toothpaste, umbrellas, and vitamin capsules.[5]

But despite the pervasive influence of petroleum products on our culture, oil and gas account for only 65 percent of the energy we use. The dirty secret of our glittering high-tech age is that we still rely very much on coal. Twelve percent of our electricity is derived from burning dirty, traditional coal. In her book *Coal: A Human History*, Barbara Freese illustrates how influential coal has been through the ages. Used by the Romans to keep warm after they invaded Britain, coal has been used as a fuel for centuries. In the 1660s, as wood in England was becoming a scarce commodity, our old friend and *Philosophical Transactions* publisher Henry Oldenburg used coal to heat his house. He might also have noted that doctors of the day recommended burning coal as a way to purify the air against the Black Death. Coal was the driving force of the industrial revolution and the mass-production society, powering James Watt's steam engines and, when baked into "coke," fueling the burgeoning iron industry. The British Empire, it's fair to say, rose to its greatest heights on the heat of burning coal.

Why are we still using it? The answer is simple: coal is abundant, cheap, and effective. For all the excitement about alternative sources of energy like nuclear, solar, wind, and gas, conventional coal is still one of the best ways to generate power. Every day when we leave our houses here in Toronto, we can smell the effect of coal in the air, which is not surprising since Ontario, one of the most technologically advanced and wealthiest places in the world, runs five coal-fired generating plants, including Nanticoke, on Lake Erie, one of the largest coal-fired stations in North America. The computer that we're using to write this book is running on electricity generated by coal. We're not as far from the world of Henry Oldenburg as we might think.

And this is exactly why ingenuity in the energy field is so

desperately needed. For all the benefits of oil, gas, and coal, there are serious drawbacks to our current energy model — drawbacks that are killing people every day. These drawbacks are not hard to discover. Simply flick on your radio during any summer day in any city in North America and within a few minutes you'll hear a perky meteorologist announce the Air Quality Index number, or the AQI. That number basically tells you if it's safe to go outside and breathe the air. The AQI is measured by taking hourly measurements of smog, which is made up of the six most common air pollutants: sulphur dioxide, ozone, nitrogen dioxide, sulphur compounds, carbon monoxide, and particulate matter. The concentration levels of these compounds are converted into a number using an index, and the pollutant with the highest number becomes the AQI reading. If the number is below 32, you're in good shape to go outside. If it's between 50 and 99, the air can have short-term adverse health effects for you and your pet, but, just so you know, significant damage can be done to vegetation at those levels. So while you might hack a bit or your kids might have a mild asthma attack, the plants in your garden are dying. An AQI reading of over 100 means that it's simply dangerous to go outside. In 2001, we had 23 smog days in Toronto.

Smog comes chiefly from the kind of energy we use, namely the oil, gas, and coal. When we burn hydrocarbons at high temperatures, like we do when we drive our cars, a whole soup of noxious chemicals is released into the air, and then, when they are combined with sunlight, they turn into smog. That we have done so little to ameliorate this problem is actually shocking, especially since the debates we're having today about pollution and health have their roots in the seventeenth century.

The ill effects of coal have of course been known for a long time. In 1661 a man named John Evelyn published a book called *Fumifugium: The Inconveniencie of the Aer And Smoak of London Dissipated together with some Remedies Humbly Proposed*. Translated from Latin, *Fumifugium* roughly means "chase away smog." The book was nothing short of an anti-pollution manifesto, with practical solutions to the problems included. With his dramatic flowing hair, prominent nose, and pencil moustache, Evelyn cut a rather dashing figure in London and he clearly knew it. The great diarist of the time, Samuel Pepys, described Evelyn as "a most excellent person he is, and must be allowed a little for conceitedness; but he may well be so, being a man so much above others."[6] Perhaps he was a little conceited, but Evelyn was no snob. Though regularly invited to sit at court and converse with King Charles II, Evelyn detested the banality and the flattery required for the occasion, preferring instead to talk about his passion: air pollution.

Reading the *Fumifugium* today you might well imagine that Evelyn was not describing the effects of coal smoke on the citizens of seventeenth-century London, but a smoggy summer day in, say, Toronto when the AQI is in the 70s and the beaches of Lake Ontario are closed due to polluted water. Almost everything that Evelyn writes about, from the "clowds of Smoake and Sulphur" which are a "hazard to health" to the "Aer that is corrupt and insinuates it self into our vital parts," are issues we're still dealing with today. Here is just one of his many descriptions of smog, with, as you will note, a very early description of acid rain:

> This horrid Smoake which fouls our Clothes and corrupts the waters so as the very Rain, and refreshing Dews which fall in the several Seasons, precipitate this impure vapor, which, with its black and tenacious quality, spots and contaminates whatsoever is expos'd to it.[7]

Evelyn goes on to describe in vivid detail the health effects of smog, how it causes the incessant fits of "Coughing and Snuffing" and "Barking and the Spitting" in public places. He even blames the smog for causing half the deaths in London, for corroding the buildings, and for polluting the water. The culprits, in his mind, are the big factories and heavy industry. Evelyn's remedy is simple: move the factories away from urban centres and then outlaw the burning of coal except in special circumstances. The ingenuity that he proposed in 1661 might well be heeded today.

But now we have an even more troubling issue to contend with, one that would have been utterly unimaginable in Evelyn's day, and that of course is climate change. According to the Intergovernmental Panel on Climate Change, carbon dioxide levels in the atmosphere have increased 30 percent from pre-industrial times and are continuing to rise. Levels of other green-house gases such as methane have also risen significantly.[8] Because climate modelling is such a complicated science, finding conclusive proof of climate change is difficult, perhaps even impossible. This has given comfort to critics of climate change, who hold that until there is conclusive proof, action is not required. But for us, the uncertainty plays the other way: the prospect of really damaging effects of climate change, even if they are only predictions and not established fact, are sufficient to cause reasonable people to take action today.

This book is about both energy and ingenuity. It's about finding ways to preserve the necessary and positive consequences of energy that will drive economic and social prosperity, and it's about finding ways to minimize energy's negative consequences: pollution, environmental degradation, climate change, and social unrest. We've asked over a dozen people who are deeply connected to these issues — financially, academically, industrially, policy-based, and journalistically — to provide their ideas and their solutions, their ingenuity, to one of society's most fundamental and most challenging problems.

We tied this project to the concept of ingenuity because of the extraordinary work of our friend Thomas Homer-Dixon, who has written the best-selling book *The Ingenuity Gap*. As a professor at the University of Toronto and the director of the Peace and Conflict Studies centre, he has spent a lot of time trying to find out how we can generate the ingenuity to solve our social, technological, and environmental problems. According to Homer-Dixon, all societies are trapped by their growing demand for ingenuity versus their ability to supply it. Societies that cannot supply enough ingenuity are vulnerable to a myriad of problems. They have what Homer-Dixon calls an ingenuity gap.

In his book Homer-Dixon describes in detail the consequences of ingenuity gaps and offers ways in which we might close them. But he also says that the problems facing us are quickly beginning to outstrip our understanding, and that we are very close to reaching a point of no return. "There is still time," he writes, "to muster that ingenuity and will, but the hour is late."[9]

If the hour is indeed late, and we believe that Homer-Dixon is right about this, then we'd better start to muster the ingenuity we're going to need in order to fuel the future. And what better person to start that journey with than Thomas Homer-Dixon himself.

A HISTORY OF ENERGY

4.5 BILLION B.C.	Energy from the sun first arrives on Earth.
560 MILLION B.C.	Plants use solar energy for photosynthesis to convert carbon dioxide and water into oxygen and carbohydrates. These organic materials eventually settle on the ground and in the water and are buried. Under deep layers of sediment, heat and pressure transform these materials into hydrocarbons, or fossil fuels.
500,000 B.C.	Humans learn to use fire that is spontaneously lit by lightning.
15,000 B.C.	Stone lanterns originally filled with animal or vegetable fats for fuel and fibre wicks are left deep in the Lascaux caves in France, well beyond the reach of sunlight.
7000 B.C.	Humans develop their own fire-making techniques using wood friction and iron pyrites and flint.
5000 B.C.	Egyptians use sails on their boats to capture wind energy and propel them along the Nile.
3000 B.C.	"Rock oil" is used by Mesopotamians in adhesives, ship caulks, medicines, and roads.
2900 B.C.	The Egyptian king Menes diverts the course of the Nile by damming it.
2600 B.C.	Sumerians use olive oil for cooking and in alabaster lamps.
2000 B.C.	Lamps using refined crude oil are developed by the Chinese for light and heat in homes. Coal is used in burning funeral pyres in Wales.
1792 B.C.	Windmills are built for irrigation by the Babylonian emperor Hammurabi.
700 B.C.	Persians grind grain with vertical-axis windmills.

▼ CONTINUED ON PAGE 26

01.

BRINGING INGENUITY TO ENERGY

THOMAS HOMER-DIXON

ENERGY IS OUR LIFEBLOOD. Without an adequate supply at the right times and places, our economy and society would grind to a halt.

Canadians are profligate users of energy; in fact, we have one of the highest per-capita rates of consumption in the world. If we were smarter about things, we would consume much less energy to support our current standard of living, and we would produce this energy with much less damage to our natural environment.

What does it mean to be smarter about things? Over the past decade I've given a lot of thought to this question. I've tried to understand why some people and societies are good at solving problems while others aren't. A key factor, I've decided, is the ability to generate and implement practical ideas. I call these practical ideas *ingenuity*, and people, organizations, and societies that can't supply enough ingenuity at the right times and places face an *ingenuity gap*.[1] This perspective helps us to understand why we're finding it so hard to change our energy habits.

Ingenuity, as I define it, consists of sets of instructions that tell us how to arrange the stuff in our world in ways that help us to achieve our goals. The sets of instructions are like cooking recipes, and these recipes allow us to manipulate, process, and reconfigure the matter around us — the materials in the ground, the gases in the atmosphere, and the organic components of our biosphere — into things that improve our lives.

Take, for example, the laptop that I'm working on at the moment. By itself, this machine probably has more power than all the computers available to the U.S. Department of Defense in the 1960s taken together. Yet this device consists of nothing more than reconfigured rock and hydrocarbons. We have extracted materials from the ground and, by following an immensely long and elaborate set of instructions, refined and manipulated those materials into the remarkable object sitting in front of me. The same is true for every human-made thing around us, including the lights above our heads, the furniture we're sitting on, and the food on our plates.

If *technical ingenuity* consists of recipes for reconfiguring matter to make technologies like laptops, cars, and furniture, then *social ingenuity* consists of recipes for arranging people to form key organizations and institutions like court systems, markets, and parliamentary democracies. Although ideas for new technologies tend to attract most of our attention, it turns out that social ingenuity is more important. In fact, social ingenuity is a prerequisite for technical ingenuity. We don't get the new technologies we want unless our economic institutions — especially our markets — reward innovators for the risks they take; and well-functioning markets take huge amounts of ingenuity to design, set up, and run.

This ingenuity model subsumes, and is ultimately far more powerful than, the conventional neoclassical economic model that dominates policy discussion surrounding critical issues like energy. Within the economic model, human beings are defined as rational maximizers of their well-being. The model's keystone concepts — concepts that we see every day in the business pages of our newspapers — are consumption, production, investment, and savings. And the model's system inputs — what economists call factors of production — are capital and labour. (For economists, capital consists mainly of the machines used to make things like cars or laptops, while labour is the work applied to running these machines.) In general, conventional economists give little thought to the independent productive role of ideas.

In contrast, within the ingenuity model, human beings are defined as pragmatic problem solvers. The model's keystone concepts are ingenuity requirement and supply. And the model's system inputs — or what we might call factors of problem solving

OUR MOST IMPORTANT CHARACTERISTIC, AND THE THING THAT TRULY DISTINGUISHES US FROM OTHER SPECIES, IS OUR CAPACITY TO GENERATE AND IMPLEMENT IDEAS TO SOLVE OUR PROBLEMS.

— are ideas, energy, and matter. Human beings use their *ideas* to guide the application of their *energy* to reconfigure the *matter* around them into the things that meet their needs. Within the economic model it sometimes seems that human beings are little more than walking appetites. According to the ingenuity model, however, our most important characteristic, and the thing that truly distinguishes us from other species, is our capacity to generate and implement ideas to solve our problems.

How can we use the ingenuity model to better understand the problems of energy consumption and production in Canada? For one thing, this model disciplines us to focus on two separate issues: the factors that boost our requirement for energy-related ingenuity, and the factors that hinder our ability to supply the needed ingenuity at the right times and places. The model also encourages us to recognize that the really big obstacles we face are social, not technical.

In the remainder of this brief chapter, I'm going to focus on why it's so hard for us to make the transition to "green" energy production, including solar, wind, micro-hydro, and landfill methane power generation. The obstacles to green energy in Canada are mainly social, and the ingenuity we must supply to overcome them is also therefore mainly social.

Let's begin with the economic obstacles to green energy. I believe that there are two main ones: first, our energy prices don't reflect energy's true costs, and second, we need high consumption to sustain our economy. The first is a tough problem, but it's potentially solvable within the context of our current economic system. The second is far more intractable, because it goes to the very heart of the way modern capitalist economies function.

Green energy could compete with conventional energy — from sources such as oil, natural gas, coal, hydro, and nuclear — if the prices we paid for conventional energy accurately reflected the full costs of its production and use. At the moment, our society, the natural environment, and future generations are all providing huge subsidies to conventional energy. As Joyce McLean of Toronto Hydro notes, "The crux here is that people are not paying the true cost of their [energy] bill. Green power, with no hidden costs, is far from being more expensive. But that's confusing to the average person."[2]

Economists use the term "negative externalities" to refer to costs not included in a good's price because they are borne by people not directly involved in producing, buying, selling, or using the good. For example, the prices of gasoline and electricity in Ontario don't reflect the cost to public health of air pollution produced when people burn gasoline to run their cars or fossil fuels to produce electricity. Similarly, these prices don't remotely reflect the likely cost of the climate change caused by the carbon dioxide emitted when we burn gasoline or fossil fuels.

These external costs of energy generation are very high. Public health officials estimate that air pollution in Toronto alone causes "between 730 and 1,400 premature deaths, and between 3,300 and 7,600 hospital admissions each year." They go on:

These premature mortality and hospitalization estimates, while significant, greatly underestimate illness associated with poor air quality in Toronto. For the last 15 years, it has been well recognized that air pollution produces a "pyramid" of health effects, with the relatively rare but more serious health outcomes (such as premature deaths and hospitalizations) at the peak of the pyramid, and the less but more numerous health outcomes such as asthma symptom days and respiratory infections (such as pneumonia) appearing in progressive layers below that peak.[3]

A major study released in July 2003 by the International Institute for Sustainable Development in Winnipeg puts numbers on some of these external costs. The results are impressive and important. The study used advanced methods of "full-cost

THE FULL COST OF PRODUCING ELECTRICITY BY BURNING COAL IS ACTUALLY 50 PERCENT HIGHER THAN THE CURRENT MARKET COST.

accounting" to estimate the health and climate-change costs per kilowatt-hour of electricity generated in eastern Canada's thermal power plants. Even using a relatively limited tally of costs (since many of pollution's effects on health are almost impossible to estimate), the study showed that the full cost of producing electricity by burning coal is actually 50 percent higher than the current market cost.[4]

Much of this cost will be borne by our children and grandchildren in the form of poorer health and a damaged natural environment. This is a subsidy paid across time — paid by generations in the future to us in the present. Conventional energy is also heavily subsidized across economic sectors and geographic locations. For example, the Ontario public have been saddled with an immense debt from the construction of nuclear reactors, and will be paying that debt through their Hydro bills every month *for decades*. And throughout Canada, urban residents are subsidizing the provision and maintenance of infrastructure, including energy infrastructure, in surrounding suburbs.

Not all subsidies are bad. Sometimes they're needed to balance social fairness against economic efficiency. But the subsidies provided by hidden external costs produce all kinds of nasty results. When people don't pay the full costs of the production and use of a good, they have an incentive to waste it and a disincentive to apply ingenuity to conserve it. If conventional energy were properly priced, it would be far more expensive, and we would see a dramatic increase in the flow of ingenuity to conserve this energy and find alternatives to it. The short-term adjustment would be harsh — there would be major economic dislocations — but, in the medium to long term, Canadian society could be much wealthier, because all that new ingenuity would make our economy vastly more efficient and technically advanced. We could also sell this energy-related ingenuity around the world.

Users of conventional energy are, in a sense, the biggest welfare bums in town. The corporations that produce this energy, and the industries and consumers who burn it, which means every one of us, are all on the dole. It's time to start paying our way.

We might be able to tackle the deeply rooted vested interests that benefit from, and rely upon, these subsidies. It's just possible that, through an immense effort of political and social will, we could reform our energy markets sufficiently to do the job. Unfortunately, though, we would still have to deal with the second economic obstacle: healthy capitalist economies rely upon ferociously high rates of consumption of goods and services, and this reliance tends to discourage a transition to a green economy whose principal aim is to lower the throughput of energy in the economy.

Understanding this obstacle requires a short digression on the nature of modern capitalism. Competition among firms encourages relentless technological innovation, which in turn steadily boosts labour productivity: as companies try to win in a Darwinian marketplace, they often replace relatively expensive labour with new technologies. This means that an individual worker can produce more for a company, but it also means that redundant workers are laid off. In order to prevent this pool of technologically displaced labour from becoming too large and both a drag on the economy and a source of social unrest, new jobs must be created through economic growth. In other words, as some sectors of the economy use less and less labour, new sectors and enterprises must be created to absorb the displaced labour.

If these new sectors and enterprises are to prosper, there must be sufficient economic demand for their products. Capitalist societies are therefore constantly engaged in demand creation. They must socialize their citizens to be insatiable consumers (the "walking appetites" of the neoclassical economic model discussed above). Advanced capitalism can only survive if it generates constantly rising material expectations and, in turn, chronic material discontent within the economically active population, despite increasing material abundance.

A large body of research shows that, beyond a modest threshold of per-capita income (less than one-third of the level of per-capita income in advanced economies), the correlation between greater wealth and greater happiness breaks down.[5] If so, why are we — and our elites, policy-makers, and governments — so obsessed with sustaining economic growth and generating endlessly greater wealth per capita? It's as if we've become addicted to buying things. The act doesn't really make us happy, except for a moment after the purchase, but we keep doing it over and over again anyway.

Western societies, and increasingly our global society as a whole, have locked themselves into an economic and social system that can remain stable only through endless growth. Without such growth, and the constantly rising personal consumption that accompanies it, new jobs won't be created, economic demand will stagnate, and deflation will set in. (Note that the world economy is currently struggling with exactly this problem: enormous excess productive capacity, vast pools of surplus capital and labour, insufficient demand to absorb these factors of production, and, as a result, incipient deflation.) Moreover, the distributional struggles between rich and poor — struggles ever-present beneath the surface of societies with highly unequal distributions of wealth and power — could tear our societies apart. Without the cultivation of insatiable desires, some technologically displaced labour will remain unemployed, which over time could produce a powerfully aggrieved and potentially revolutionary underclass.

In this macroeconomic environment, appeals for energy conservation must climb a very steep hill. Conservation has a pejorative "eat your peas" connotation. The not-so-subtle implication is that it's bad for the economy — and almost un-Canadian. We need to consume in order to grow, and we need to grow in order to be happy. Dealing with this problem will require advocates of green energy to revisit some fundamental issues about economic growth and the nature of capitalist economies that they've been reluctant to address since the aborted "limits to growth" debate of the 1970s.

If these economic obstacles to green energy are formidable, so are the political obstacles. Again I see two. First, the political and bureaucratic systems that form the environment within which advocates of green energy must press their case, and that generate the regulations governing green energy's deployment, are often hopelessly cumbersome and inefficient. Journalist Gordon Laird writes, for example, about the Toronto Renewable Energy Cooperative's attempts to get approval for a wind turbine at Ashbridge's Bay, east of Toronto:

> There are three levels of government, technical and compatibility problems, city bylaws that need be rewritten or revised, and a host of other complications. The growing list of government jurisdictions and departments requiring reports and approval is considerable — City of Toronto, Navigation Canada, Transport Canada, Toronto airport, Committee of Adjustment — and then there is the federal environmental assessment, a process that requires its own small mountain of paperwork and consultations. There's also a list of unwritten "unofficial approvals . . . wink-wink approvals you better have."[6]

Moreover, as was shown years ago by the economist Mancur Olson, bureaucracies, political systems, and economies become more complex and rigid — even sclerotic — over time as they add layer upon layer of law and regulation, and as they accumulate competing and overlapping centres of authority. As years go by, powerful vested interests establish themselves in the niches of these systems and do everything they can to prevent change that would affect their interests.

Energy supply and use affect every person, group, and segment of society, so everybody has a powerful interest in these matters. In such an environment, it's very difficult to reach a political consensus, and it's very easy for powerful interest groups to derail reforms.

The second political obstacle arises, somewhat paradoxically, from green energy's greatest strength — its emphasis on small-scale, local, and decentralized energy production. Its advocates tend to be small, community-based, and loosely networked groups, with commensurately little political clout. On the other hand, large, centralized power projects like nuclear stations, massive fossil fuel generating plants, and hydroelectric dams depend on huge corporations and government bureaucracies to build, run, and maintain them. These corporations and bureaucracies become concentrations of political and economic power in Canadian society, and too often they become immense and focused vested interests opposed to real reform of our energy practices. Green energy's advocates, in contrast, are often underfinanced, poorly co-ordinated, and ideologically diffuse. In the fierce political struggle over energy policy, they can't hope to win against very powerful, very rich, and very entrenched vested interests.

The four obstacles to green energy that I've identified here — two economic and two political — are formidable barriers to change. Most fundamentally, they are serious obstacles to the supply of the ingenuity we need to solve the energy problems we face. There are things, however, that we can do. Canadians can lobby their governments to adjust energy markets so that prices better reflect energy's true costs. They can work to streamline the political and bureaucratic decision-making processes that affect new energy policies. And green-energy advocates can make their political action more co-ordinated and coherent, in order to counterbalance the dead weight of vested interests behind conventional energy.

But as Canadians we also need to reflect much more on our broader context of institutions, incentives, and values — a context that leads us, too often, to choose a mode of living that requires huge amounts of energy over one that treads more lightly on Earth.

A HISTORY OF ENERGY (CONTINUED)

500 B.C.	The Greek philosopher Heraclitus describes a world in a perpetual state of flux of opposites, particularly between earth, fire, and water. Thales, another Greek philosopher, discovers static electricity when he observes that rubbed amber attracts certain materials. The word electricity will derive from the Greek word for amber, *elektron*.
260 B.C.	Archimedes invents the Archimedean screw, a helix-shaped screw in a tube, used for raising water.
212 B.C.	Archimedes designs a way to defend the harbour of Syracuse using solar energy, by reflecting a concentrated beam of sunlight in giant mirrors to set fire to Roman ships.
200 B.C.	Hero of Alexandria describes the first known wind device and the first steam engine, called the *aeolipile* in his writings.
600–900 A.D.	Windmills for irrigating crops appear in Persia (near present-day Iran and Afghanistan). Still to the present day, some of the world's oldest windmills will be located in Afghanistan.
700–1100 A.D.	Wind-powered Viking ships rule the North Atlantic.
1000 A.D.	By this date the Chinese are using gunpowder, a mixture of sulphur, saltpetre, and charcoal, for fireworks and signals.
1228 A.D.	The first shipment of coal from Fife and Northumberland lands in London. These broken lumps from underwater outcroppings are given the name sea coal.
1264 A.D.	Marco Polo visits the Persian city of Baku (in present-day Azerbaijan), on the Caspian Sea, and sees oil as it is collected from seeps. He will later write about people here using oil for medicinal purposes and the burning of petroleum at a "fire-temple."
1610 A.D.	A quarter of all English maritime trade (by weight) is coal from the North, three times the size of all other coastal shipping. The proportion of trade will rise from 25 to 40 percent by 1660.

▼ CONTINUED ON PAGE 54

02.

AT THE FRONTIER OF ENERGY

GORDON LAIRD

EDITORS' INTRODUCTION

Thomas Homer-Dixon's discussion of energy and ingenuity raises some challenging questions. How is it possible to change something as dominant and influential as the current energy industry? Aren't we all virtually powerless to affect such an enormous infrastructure? To answer these questions, it's worth remembering the work of another great Canadian thinker, Harold Innis.

An acclaimed communications theorist and the precursor to Marshall McLuhan, Innis was deeply interested in how powerful organizations or empires are altered from the margins. Innis believed that greater experimentation and more radical ingenuity is acceptable in out-of-the-way places, where the risks are lower and authoritative interest groups have less influence. This provides the space required for new social forms, such as innovative mechanisms of communication or media, to develop. These new social developments in turn encourage those on the outside to develop positions of power that can eventually challenge the authority at the centre. This process can result in real change — or, to use a more in-vogue term, a paradigm shift. As a result, the best way to see the future of an empire is to look at what the barbarians at the gate are doing.

The energy world is no different than most other empires, except that the frontiers are wilder and the experiments are much riskier. While most of us enjoy the comforts of easily accessible, cheap energy, what we don't see is how much effort, hazard, and pain it costs to keep that energy flowing. The multi-billion-dollar campaign is being waged on the forsaken frontiers of the world, often in the land of a people many of us have never heard of, where the enemy might just as likely be a group of saboteurs or the fierce elements of the Northern tundra.

Gordon Laird, author of *Power: Journeys Across an Energy Nation*, the best-selling book about how energy has shaped Canada, travelled to several remote locations to see what is brewing at the margins of the conventional energy empire. Here is Laird's report from the front lines.

SOME THIRTY YEARS AFTER they were banished from Canada's Western Arctic, pipeline workers are back again. Pushed out by intense Aboriginal lobbying and the Berger Royal Commission of 1977, which concluded that the Mackenzie Valley wasn't ready for a massive energy boom, drillers, seismic crews, and company officials are now returning to this huge span of river, tundra, and boreal forest. At the very top of the Mackenzie — the very edge of Canada's northwestern frontier — lie some 64 trillion cubic feet (1.7 trillion cubic metres) of recoverable natural gas reserves, with untold trillions more possibly hidden along the pipeline route, a rich sedimentary basin that runs all the way from Alberta to the Arctic Ocean.[1]

From the Alberta border north, the Mackenzie River snakes through mountains, boreal forests, and modest Aboriginal communities. As it approaches the Arctic Ocean, the Mackenzie sprawls sideways and becomes the world's tenth-largest river delta — a flat expanse surrounded by endless tundra and scrub. Beneath this land at the top of the Mackenzie, some of the richest gas plays on the continent lie under caribou wintering grounds.

It is for this prize that the pipeline workers return, driven by continental markets anxious about looming natural gas shortages. And, incredibly, the region known for one of the biggest industrial flops in living memory — the stillborn pipeline of the

1970s was just one aspect of Pierre Trudeau's failed Arctic energy plan — could be shipping as much as 2 billion cubic feet (56 million cubic metres) of natural gas daily as early as 2008.

How did this happen? Through some 180-degree turn of history, you are now just as likely to see Inuvialuit, Gwitch'in, or Sathu running business meetings, cutting deals, and working the drillfloor. Indeed, the First Nations of the Mackenzie are one-third owners of what will become Canada's first Arctic mega-project, right alongside corporate majors like Imperial, Shell, Conoco, and ExxonMobil, all of whom hold substantial reserves from the Arctic's first wave of gas exploration. The result will be the continent's single longest pipeline and the first megaproject of the twenty-first century: a 1,350-kilometre-long string of steel that will carry a twenty-year supply of natural gas to hungry North American markets.[2]

It's a long way from the days when Aboriginal leaders accused companies of genocide and colonization. "We're business-oriented people around here now," explains Roy "Bunker" Wilson, an Inuvialuit community liaison for Imperial Oil. "People [are] not just asking about jobs, but about ownership of those jobs. It's not just, can I go work? — it's more like, can I send my truck to work?"

We're on the land south of Inuvik, a rough Arctic boom-town that serves as control centre for all things pipeline. But out here at the treeline, overlooking the barrens to the north, a small piece of history unfolds. Two weeks before the winter ice roads close, a mobile drill unit plods along the proposed pipeline route. On the edge of the Mackenzie Delta's alluvial plain, I watch as a geotechnical team drills nine-metre core samples in the permafrost. Theirs is the mundane work of extracting soil for regulatory applications and pipeline engineering plans, all part of a $270-million project definition program launched by the Aboriginal Pipeline Group and its corporate partners.

Even to locals, the shift in perspective is nothing short of astounding, although social changes were happening well before the first Arctic energy boom of the 1970s. "It's the elders who are the most connected to the land. It wasn't until the 1950s and 1960s that we got into a cash economy," explains Bunker as we traverse the border of the Arctic barrens in a giant snowcat, not more than forty kilometres from the Mackenzie Delta. Dwarf trees scatter the land out to the south, and rolling hills, the wintering ground for caribou, run to the north. "My generation and

THE $5-BILLION PIPELINE, AND THE POLITICAL MOMENTUM BEHIND IT, SHOWS THE TRAJECTORY OF POWER IN THE NEW CENTURY.

the generation before me now live a different life." It's a common story in Canada's North: many Aboriginals want jobs, partly because traditional pursuits like hunting and trapping can't pay for groceries and rent.

The sheer scale of this project, all privately financed, underlines the importance placed on energy frontiers in the twenty-first century. The $5-billion pipeline, and the political momentum behind it, shows the trajectory of power in the new century. Considerable wealth and effort will be mobilized in order to sustain the energy norms of the previous century, even if this means laying pipe across some of the world's most forsaken terrain.

The global push to control and develop frontier resources is hardly unusual. Nations around the world are delving deeper and deeper into remote territory to locate marketable fossil fuels.

In conjunction with the world's largest corporations, govern-ments of all stripes are pushing hard to find and exploit non-renewable resources, in order to avoid a short supply of fuels. Nations are increasingly fearful of an energy crisis, a fact that's reflected in official plans to strategize "secure supplies" of fossil fuels and abundant megaprojects as we attempt to forestall the inevitable and final depletion of non-renewable energy.

A surge in new megaprojects is accelerated by a series of tectonic economic shifts that are transforming previously cheap commodities like natural gas into one of Earth's most strategic resources. A fuel like natural gas, valued for its versatility and rel-atively clean burn, is the energy source that could bridge a global shift to more diversified, sustainable forms of power — energy efficiency, renewables, hydrogen cells, and beyond. Visionaries have proclaimed natural gas as the missing link between a fossil fuel economy and the new hydrogen economy, especially with the advent of prototype (but still largely unproven) fuel-cell tech-nologies that convert natural gas directly to electricity.

Yet the reality is a lot more complicated. As old coal plants and new power stations run high-efficiency gas turbines, elec-tricity prices are increasingly tied to natural gas. And continental power demand has not been effectively managed, so that mass inefficiencies and waste inevitably drive up commodity prices. Moreover, new and unconventional sources of crude oil, such as the booming oil sands of northern Alberta, which require upwards of two times more energy than conventional produc-tion, are being developed to feed our outsized energy hunger. It is one of the largest industrial developments in the world, with some $51 billion in existing and new projects. With oil sands expected to provide half of Canada's crude output by 2005 — tapping a bitumen pool that's almost as large as the proven reserves of Saudi Arabia — the energy requirements for crude production will be profound.

Some critics have actually described oil-sands crude production as an energy upgrading process, not a new resource. Given the scale of development — oil-sands crude has already outpaced conventional output in Alberta — it is likely that Canada's production of synthetic crude will profoundly affect natural gas reserves. "There does not seem to be enough natural gas in sight to supply both the oil and the gas desired by the U.S.

THE COMPOUND EFFECT OF SPIRALLING DEMAND AND FOSSIL FUEL SCARCITY, COMBINED WITH THE VULNERABILITY OF AN ECONOMY THAT OFTEN REWARDS WASTED ENERGY, MEANS THAT WE COULD SEE POWER SHORTAGES AND AN ENERGY-DRIVEN ECONOMIC RECESSION BEFORE THE END OF THE CENTURY'S FIRST DECADE.

above Canada's internal needs," noted a 2002 study from the Hubbert Center for Petroleum Supply in Colorado.[3] By the end of the decade, it predicts, oil-sands operations are expected to consume some 25 percent of Alberta's existing gas production, about 1.3 trillion cubic feet (36 billion cubic metres) a year. "Within a few years Canada may have to choose between selling part of their natural gas vs synthetic oil to the United States."

The compound effect of spiralling demand and fossil fuel scarcity, combined with the vulnerability of an economy that often rewards wasted energy, means that we could see power shortages and an energy-driven economic recession before the end of the century's first decade. As U.S. Federal Reserve Board chair Alan Greenspan warned in June 2003, this crunch could be first precipitated by natural gas. "Demand destruction" is the new catchphrase for prohibitively high natural gas prices that would force energy-intensive businesses — manufacturing, fertilizer, and so on — into temporary shutdown or outright closure. Hence the rush to open up Canada's Arctic reserves, something that American president George W. Bush himself has taken a personal interest in championing, opposing proposed tax subsidies for a competing gas pipeline through Alaska.

In fact, most new megaprojects will service a twentieth-century industrial model designed around the presumption of abundant and affordable supplies of fossil fuel power. The inherent problem is that non-renewable energy sources are now more interdependent than ever. And as contingent fossil fuels are depleted, fears of a domino effect — a cascading market failure not too unlike the great blackout of 2003 — could precipitate an unexpected and widespread short supply of crude, electricity, and natural gas.

In other words, there is no escape from uncontrolled demand and the uncertainties of non-renewable supply. With a large portion of Mackenzie natural gas expected to power Alberta's oil sands, oil-sands producers already export half their crude to American refineries, essentially shoring up a jurisdiction that currently refuses to upgrade automotive efficiency standards. In short, North America's huge, energy-guzzling fleet of SUVs and minivans, single-handedly responsible for the worst auto-efficiency rates since the 1970s, has much to do with why Inuvialuit drillers are punching exploration gas wells across Canada's Western Arctic. And why, if current patterns hold true, there might not be much new natural gas left over to fuel more sustainable alternatives.

Out on the ice of the Mackenzie River, I happen upon a debate over renewable energy, of all things. It's the Muskrat Jamboree in Inuvik, and as locals prepare for annual dogsled races, there's a fierce discussion about the merits of wind power. "Well, we tried it, and it fell flat," says one manager from the local power utility. "And the micro-turbines on the gas pipeline don't perform to spec, either."

Up in the Arctic, energy is a constant focus. Hunters need fuel for snowmobile forays onto pack ice and tundra, animals need vast quantities of land to get enough food energy, and everyone worries about the sheer cost of conventional energy, since few people live out on the land anymore and many communities rely upon toxin-laden diesel turbines. The fierce economics of fossil fuel consumption are clear above the sixtieth parallel. Imported diesel consumes money that would be better spent on providing local health care, supporting the indigenous hunting economy, and addressing the chronic housing shortage that can be found in almost all of Canada's Northern communities.

Inuvik has been an energy centre for years. With its own natural gas sources, it has more economic resources to put towards economic development. And as I talk to more people on the ice of the Mackenzie, I find that there are alternative energy projects across the North, from faulty gas micro-turbines in Inuvik to wind power on Holman Island. Through sheer economic necessity — and meagre government grants — people are trying to launch alternative energy projects and ambitious conservation efforts on a scale that, per capita, is unheard of in the rest of Canada. Consequently, many Northerners know the advantages and disadvantages of renewables and various energy technologies because they are already using them. It turns out that there is often a steep learning curve for any renewable application, and the North is an important testing ground for real-time energy alternatives, not just fussy little demonstration projects that we often see here in the south.

Many Northern communities are already attempting ambitious efficiency and energy projects. The community of Wha Ti, near Yellowknife, has embarked on a multi-year plan, something that began with the realization that fossil fuel usage was not only poisoning residents but also dooming the local traditional economy through high energy costs. An energy audit spelled it out: ". . . with a population of about 485 people, [Wha Ti] required 887,000 litres of diesel for power and space heating in 2001. At about one dollar per litre, and with transportation expenses, total cost for these two services was well over one million dollars."[4] At just over $2,000 per resident annually, it's an expense that's double or even quadruple that of most Canadians.

A select few communities like Wha Ti are currently experimenting with wind power, micro-hydro, and passive solar heating. Wind turbines can be a tricky technology with the ice storms and extreme cold intrinsic to parts of the North. But available capital is limited for renewable projects with a long payback period, so when seismic or drilling companies come calling, many locals concern themselves with making a living before dabbling in high-end technologies.

But efforts to evolve energy production and consumption are increasingly overshadowed by the immanent megaproject, the Mackenzie Valley natural gas pipeline. Very little doubt remains as to whether this natural gas pipeline will be profitable. With off-season gas prices near double their usual rates and conventional gas reserves fluctuating across North America, there could soon be high times indeed for anyone with new supplies. Rather, the trick is parsing a tightly woven web of interests, agencies, stakeholders, and, if those weren't enough, the looming spectre of competing energy megaprojects — largely fossil fuel — that are being launched around the world in anticipation of short supplies and erratic, dwindling reserves.

The looming change goes well beyond demand for jobs, however. Like the Inuvialuit to the north and the Sathu to the south, the Gwitch'in people of the Mackenzie have settled their land claims and now enjoy an increasing degree of self-government.

If you want to do business on their land — say, drill an exploration well or lay 1,200 kilometres of pipe — a certain degree of partnership, not to mention respect for the land itself, is now expected. Incredibly, Imperial Oil, the lead partner in the consortium with the Aboriginal Pipeline Group, actually refuses to build a pipeline without First Nations partnership. In other words, Imperial, Shell, and their partners would not build even if the Canadian federal government bulldozed local process and handed them an open ticket. "We will not proceed in the absence of the Aboriginal Pipeline Group," says Dee Brandes, Imperial's Consultation and Community Affairs manager. "If the region is opposed to the project, we're not going there."

This too is a change in perspective. Imperial, like many other companies, once took a dim view of local partnerships and control, leaving behind a surprising amount of bad feeling and mistrust along the Mackenzie. Back in the 1970s, there probably wouldn't have been Gwitch'in environmental monitors or subcontractors. Their labour might have been menial, possibly token. And today, as we survey the geotech crew who are drilling back down into another section of permafrost, Brandes concedes that it's been a challenge forging a twenty-first-century energy alliance.

On the day that I arrive on the pipeline route, just south of Inuvik, Imperial's geotech crew is about four months late for work, simply because someone botched a land-use permit application. Environmental regulators associated with Aboriginal self-government appear to be more stringent than conventional government regulatory process. An Aboriginal official at the Gwitch'in environmental authority concurs. "Anything associated with the word 'pipeline' in the NWT everybody puts under a microscope," says Darren Campbell of the Gwitch'in Land and Water Board. "And because of that, every regulatory agency is taking the proper steps." He continues, "There is a big difference in how things work here and how they work back in Alberta.

There is some head-butting involved. The best thing they did was to come in and meet face-to-face."

The local regulatory process — one that's inevitably tied to federal National Energy Board deliberations — is stringent largely because of the growing scope of self-government. "A lot of organizations here are not top-down process," Campbell says. "The chief of any given community takes direction from the population. So you need meetings, and that's not always so easy." Paradoxically, self-government in the Mackenzie also underpins the new business spirit of the region. It's precisely because Aboriginals have control that development is permitted and promoted. What this all means for the first megaproject of the twenty-first century is that the task of forging partnerships is potentially a bigger job than building the pipeline itself. The megaproject is the relationship, and that mantra shapes everything that happens along the Mackenzie.

The Mackenzie Valley pipeline will also be Canada's first megaproject launched entirely on market criteria, with no subsidies, grants, loans, kickbacks, or public boodle of any kind. It contrasts sharply with the Arctic program of the Trudeau era, which sometimes matched private money dollar for dollar, or even with the Alberta oil-sands development in recent years, where federal tax subsidies for "mining" dolled out millions to companies that invested in the processing and refining of synthetic crude. No, this pipeline will be an investors' pipeline — a scenario that suits many Northerners just fine, given the spotty historical record of subsidized industry in the North.

It's been a long economic struggle to gain land claims and set up local self-government. People like Premier Stephen Kakfwi, who was a radical opponent of the pipeline in the 1970s, now understand the future to be in compressor stations, high-tech Arctic drill rigs, and Aboriginal entrepreneurs. He explains that everything from the last thirty years has pointed towards this moment when Aboriginals can bargain as equals, and it all stems from the valid recognition of Aboriginal title, self-government, and local control. "You're on the side of the angels when you say, this is my land," he says. "Anybody would die for an opportunity to be able to say those things. I've spent my whole life trying to make it happen. It's wonderful, because you know it's possible."

A new North is emerging, and it's not like anything else we've ever seen. This pipeline is more than another large piece of infrastructure; it is the manifestation of a new political and economic zone with its own rules and clout, something quite unlike the bloodied energy plays of the developing world. Up here, locals often call the shots. So if it's gas you want, you must first reckon with the Mackenzie.

But as locals engage with a megaproject, talk of renewables and energy alternatives seems to fade. Governments themselves are even more conflicted. Non-renewable resources are big business, yet their non-renewable nature poses a profound problem for economies that become too dependent on them. In the Northwest Territories, for example, almost 50 percent of all funding for resource development subsidizes the oil and gas industry through road construction, whereas only 10 percent is devoted to economic diversification, social impacts, and environmental management.[5]

In other words, the Mackenzie pipeline is a massive social, economic, and environmental experiment: Can a megaproject serve everyone? Despite the promise of today's Aboriginal–corporate partnership, the long-term trend for Northerners is ultimately unclear, since there have been few studies of the cumulative impacts of multiple energy megaprojects and the outcome of economic dependence on non-renewable resources. One 2001 pilot study by the United Nations Environmental Program (UNEP) concluded that up to 80 percent of the Arctic will be affected by mining, oil and gas exploration, ports, roads, and other developments by 2050 if development continues at its current rate.[6] "At the turn of this new millennium less than 15 percent of the Arctic's land was heavily impacted by human activity and infrastructure," says Klaus Töpfer, Executive Director of UNEP. "However, if exploration for oil, gas, and minerals [and] developments such as hydro-electric schemes and timber extraction continue at current rates, more than half of the Arctic will be seriously threatened in less than fifty years."

Thanks to climate change, energy alternatives, and megaprojects, Canada's North is on the cutting edge of our energy future. But in many ways, all North Americans are caught between the

FANTASTIC AND SOMETIMES HORRIBLE THINGS HAPPEN OUT ON THE WORLD'S BORDERLANDS — EVERYTHING FROM TORTURE IN CENTRAL ASIA TO PIPELINE SABOTAGE IN NIGERIA — JUST SO THAT WE CAN MAINTAIN THE NORMALCY OF OUR ENERGY NETWORKS.

seductive lure of energy megaprojects, the dire need for energy conservation and alternative generation, and the economic and environmental consequences that loom in the future.

It is at the edge of our energy kingdom that we attempt to reconcile the inevitable decline of fossil fuels with our deep dependence on fossil-fired economies. How deep will we drill? How far will we travel? How much will we pay to maintain the past? Fantastic and sometimes horrible things happen out on the world's borderlands — everything from torture in Central Asia to pipeline sabotage in Nigeria — just so that we can maintain the normalcy of our energy networks. But lasting solutions based on a model of non-renewable resource extraction are, by definition, virtually impossible, so the ends of our engineered world continue to be transitory places, prone to extremes. These are places that manifest new innovations — such as cutting-edge energy efficiency technology already deployed in the oil sands — as well as instability, lawlessness, and poverty.

For example, China has begun its own massive, $24.5-billion pipeline, stretching 3,700 kilometres eastward across the country from Urumqi, the capital of Xinjiang Province (East Turkestan), to Shanghai, simply to supply natural gas into the polluted epicentre of the Yangze economic boom.[7] The Central Asian frontier is crucial to China's economic survival plans, notwithstanding sporadic efforts by Uighur rebels to push colonists back across the Great Wall. Xinjiang has an estimated 36 trillion cubic feet (one trillion cubic metres) of recoverable natural gas buried deep beneath the sands of the Taklamakan Desert, plus the largest single source of crude oil in western China. By virtue of geology, globalization, and just plain bad luck, Uighur nationalism — and its energy-rich homeland — stands in the way of the planet's most voracious consumers of power.

Consequently, surreal and sometimes brutal scenes play out here daily, a distant link in an ever-globalizing cycle of resource extraction, manufacture, and consumption.

There are many mysteries in Xinjiang, but perhaps the most unlikely one is this: Why does its capital, Urumqi, one of the most landlocked cities in the world, have so many seafood restaurants? Fresh off the plane from Beijing, this was the first question that vexed me. Not Xinjiang's simmering religious wars, nor China's western gulag with its political prisoners, nor the vast Taklamakan Desert directly to the south, cluttered with drilling rigs that plumb the vast oil and gas deposits. No, I was puzzled by the squirming eels from Guangzhou.

Wandering down Xinhua Street, past Urumqi's Holiday Inn, I'm 2,400 kilometres from the nearest ocean and all around me are seafood eateries, whose billboards and owners excitedly proffer overpriced meals from the deep. This former Silk Road outpost has, over the last decade, been transformed into a major Chinese city under a state-sponsored energy boom and mass resettlement program. Under the shadow of a giant Ferris wheel in nearby Hongshan Park, live lobsters peek out from dingy little tanks, and buckets of iced mussels sit beside halibut and squid. I've literally stumbled across a little slice of south China at the edge of Central Asia: cursing drivers, blaring pop music, and the quick promise of good eating.

Of course, the eels and lobsters are foot soldiers in a slow-burning battle between Chinese newcomers and resident Uighurs, Xinjiang's Islamic majority. Urumqi itself is a city so divided by money, power, and religion that even food has become political. Out here on the edge of Central Asia, imported seafood is a compelling sign of Chinese dominance, tacit proof of residency in a region that can hardly agree on a time zone, let alone mend fences in a struggle that's been simmering since the first century A.D. Here, Chinese entrepreneurs serve up seafood in air-conditioned splendour while Uighur vendors peddle kebabs and nan flatbread from carts and makeshift market stalls.

The seafood is a victory sign. The Chinese have largely won this war, and their conquest has much to do with the scarcity of cheap, accessible fossil fuel around the world, something that's begun to affect the economic powerhouse that constitutes China's southerly provinces and eastern coastline. Indeed, China has achieved unprecedented growth, averaging 7 percent annually for the last twenty years, thanks to the sheer abundance of two main factors: cheap labour and cheap energy. And it has been much to our benefit, the consumers of North America, that failed revolution (cheap labour) and non-renewable energy (from coal, oil, and natural gas) have been mobilized on such a grand scale. This is how China has come to supply stores like Wal-Mart with most of their product. Shenzhen, a rough economic zone that runs along Hong Kong's northern border, alone manufactures an estimated 70 percent of the world's toys.[8]

The inevitable and predictable consequence of building a modern-day manufacturing empire is a deep hunger for resources. In the case of Xinjiang, the old geopolitics of Maoism — the manifest destiny of an ideological state — has been replaced by a market-driven oligarchy deeply in need of large quantities of secure power. Because if China cannot feed its factories, refineries, and booming cities, its growth stalls and the delicate political balance of its post-communist system is potentially outpaced by rural poverty, regional upheaval, and the debilitating cost of offshore energy imports. In North America, imported crude oil is still accepted as a necessary evil, even if it indirectly funds terrorism. In China, the prospect of escalating energy costs and excessive imports is seen as the end of the miracle boom — and the onset of greater troubles.

In the midst of a state-sponsored energy boom, Uighurs have become a minority in their own capital city. The flood of Han settlers, businesses, industrial development, and police into Urumqi has, in turn, fueled an anti-Chinese campaign that has given Beijing serious cause for concern. Political bombings, assassinations, and weapons smuggling continue unabated. I've arrived in Urumqi — or Urumüqi, as the Chinese call it — only a few months after the last major clash between police and Uighur nationalists.

It's always been a complicated place. Xinjiang is the site of one of Asia's longest-running intra-national conflicts, a two-thousand-year battle over a desert kingdom. This conflict rarely captures international headlines, but open warfare, summary executions, and assassinations are commonplace. Consequently, the region is a major concern for international human rights groups like Amnesty International, who decry "gross and systematic" abuses here.[9]

Tightly controlled by the Chinese government, Xinjiang is all but invisible to the rest of the world, off-limits to foreign journalists — I'm here on a tourist visa — but not to major multinational oil companies like BP and Shell, who fly their flags up and down Urumqi's Beijing Avenue. Tremendous wealth is being taken from the ground beneath these old Silk Road outposts in the Taklamakan Desert, a region formerly valued for its cotton and strategic proximity to Russia, yet little affluence has reached the Islamic majority. Most Uighurs, Kazakhs, and other indigenous Central Asians still subsist on less than $270 annually.

China has nevertheless pledged to modernize Xinjiang for its own good with a multi-billion-dollar plan that involves importing vast quantities of bureaucrats, soldiers, gas pipeline, and squid. When the pipeline reaches full operational status in 2007, it will carry as much as 430 billion cubic feet (12 billion cubic metres) of natural gas a year.[10] Only the Three Gorges Dam project, which is costing China's government about $34 billion, is more expensive.

Local Chinese have been, predictably, the first to enjoy Xinjiang's boom, something that only fuels the longstanding feud. "The Chinese get all the favours and preferential treatment," complains my newfound host, Rebiya. "There are few jobs for we Uighurs." For emphasis, she points her finger like a gun and fires at a Chinese couple strolling by.

As we stroll along, Urumqi presents itself as two cities: a hastily built Chinese metropolis of concrete, glass, and brick, and a Uighur stronghold full of ancient mosques and markets. We walk towards the Sanxihangzi Market in the centre of town, passing the famous White Mosque, a 200-year-old institution that, she says, is a frequent haunt of the secret police. Although Sanxihangzi is the oldest market in Urumqi, Guangdong developers recently gained permission to tear down its northern corner for a gleaming high-rise. Fourteen stories high, its footprint has pushed aside fruit vendors and bread peddlers — much to the indignation of locals, who consider the market a treasure of Uighur culture.

On the surface, the market is chaotic. Pakistani traders wave bolts of imported silk, while cartloads of nan bread — salted

THIS IS A PLACE THAT STILL CONSIDERS FIREWOOD AS A CUTTING-EDGE FUEL; THE EARTHEN OVENS THAT COOK THE NAN BREAD DON'T REALLY WORK WITH ANYTHING ELSE.

and savoury — are pushed through the crowd. As we walk through the narrow maze of stalls, Rebiya picks out the most choice dried apricots as a gift, along with some sweet melons from a tall pile on a cart fresh from the countryside. This is a place that still considers firewood as a cutting-edge fuel; the earthen ovens that cook the nan bread don't really work with anything else.

By contrast, China operates on the Western mode of ever-increasing consumption; energy conservation is often a secondary concern to adding new megawatts and gigajoules at almost any price. In recent years, China's per-capita annual energy consumption was the equivalent of less than 1.2 tonnes of standard coal, less than a half of the world's average. But the International Energy Agency predicts that China's energy demand will grow by 2.7 percent a year until 2030, when annual demand will reach 3.19 billion tonnes of standard coal, almost three times current rates.[11] In 2002 alone, China's top passenger-car-maker, Volkswagen, posted a 40-percent jump in sales, and gasoline usage is expected to climb 4 percent a year until 2005. (The environmental benefit of the natural gas pipeline will come from a reduction in demand for other fossil fuels; even a small offset in

China will amount to a massive reduction in greenhouse gas emissions.)

Xinjiang's invisibility to the outside world has not happened by mistake, and the intra-national strife in Xinjiang, which exceeds even Tibet in the frequency of its summary executions, is something that China's affluent trading partners have chosen to ignore. For the first time in years, for example, China was taken off the United Nations censure list in 2002 because member nations elected not to question ongoing human rights violations. Consequently, Xinjiang's economic divide and sustained political crackdown have pushed the traditionally moderate local population towards fundamentalism, now actively exported by radicals in Pakistan, Afghanistan, and the former Soviet republics of Central Asia. Between 30 and 300 Uighur Taliban fighters were reported captured by American forces during the 2002 Afghanistan war.[12] Xinjiang's long, porous border ensures that radical elements, as well as guns and ammunition, continue to trade, despite Chinese efforts.

A large part of the energy war is diplomacy and propaganda. The attacks of September 11, 2001, presented a tremendous political opportunity for Beijing to clamp down on the region and push ahead with the pipeline plans. One government report proclaimed, "East Turkestan, An Integral Part of Bin Laden's Terrorist Forces." Official sources have claimed that Xinjiang separatists committed more than 200 acts of terrorism in China between 1990 and 2002, killing 162 people and injuring more than 440.[13] Xinjiang police claim that there were 800 "separatist incidents" in the first eight months of 2001 alone.[14] In 2002, following the war in Afghanistan, the Chinese government gained the cautious support of the Bush administration in adding an outlawed Uighur political party, the East Turkestan Islamic Movement, to its list of known terrorist groups. Later in 2002, China's Foreign Ministry reported that U.S. President

Bush and Chinese President Jiang Zemin had agreed that "Chechnya terrorist forces and East Turkestan terrorist forces are part of the international terrorist forces, which must be firmly stopped and rebuffed."

Nevertheless, the real threat of terrorism in Xinjiang remains somewhat limited. "Underground organizations that flourished a decade ago have been destroyed," *Time* magazine reported in 2002. ". . . The only known 'separatist attack' launched since Sept. 11 took place Jan. 1 when an unemployed Uighur sneaked onstage during a New Year's Day song-and-dance revue for the province's top leaders and declaimed an

IT IS HOPED THAT ASIA'S LONGEST PIPELINE WILL SOMEHOW BREAK THE STRIFE OF XINJIANG BY OFFERING AFFLUENCE TO THE INDIGENOUS POPULATION, JUST AS STRIFE IN CENTRAL TIBET HAS BEEN TEMPORARILY QUELLED BY THE EMERGENCE OF A MORE AFFLUENT AND MODERATE MIDDLE CLASS IN LHASA.

independence-minded poem."[15] Expatriate political networks have been subdued, transborder traffic in dissidents has been slowed, clerics are forced into political re-education, birth control is heavily enforced, and even the Uighur language, a distant relative of Turkish, has been banned from universities and schools for communist cadres. In short, today's Xinjiang bears a harsh resemblance to the China of Mao Zedong's cultural revolution — a forgotten zone that appears to be moving backwards in time.

It is hoped that Asia's longest pipeline will somehow break the strife of Xinjiang by offering affluence to the indigenous population, just as strife in central Tibet has been temporarily quelled by the emergence of a more affluent and moderate middle class in Lhasa. Yet the paradox is that many Uighurs resent the sheer fact of Chinese occupation, and the export of valuable resources only exacerbates this. As with Canada's Arctic pipeline, whether this megaproject will actually settle intrinsic underdevelopment and structural inequalities isn't entirely clear.

But while Mackenzie Aboriginals gain greater autonomy and self-government, Chinese efforts to control and colonize Xinjiang have intensified. "With construction on the vital West-East gas pipeline just beginning and energy investment into Xinjiang on the rise, the last thing Beijing wants to see is a resurgence of the separatist-linked violence that hit the region from 1996 to 1998," noted U.S. strategic analysts StratFor in a 2002 report. "Ironically, the same event that may force Uighur militants abroad to return to Xinjiang — the U.S. war against terrorism — at first appeared to offer the perfect opportunity for Beijing to intensify its crackdown in the region without endangering its international image."[16]

Indeed, anger against China is increasingly being redirected at the United States and its allies. "There's real disillusionment and growing anger, not so much against China, but against the United States. Mosques are now raising warriors to go and fight," said Dru C. Gladney, a Xinjiang expert at the University of Hawaii. Speaking to a Beijing audience in 2003, he reported that American policy in the Middle East is radicalizing the political underground in Xinjiang, which in turn may disable China's long-term ability to export cheap consumer items to the West. "There is no love or sympathy for Saddam Hussein [in Xinjiang] but there's certainly a lot more awareness of international politics and they're certainly in touch with Muslims around the world."[17]

China's corporate partners, led by Shell, include a collection of American and Russian multinationals. But the pipeline itself, as large and expensive as it is, offers only meagre profits, if any. If payback for the pipeline is decades in the future, if ever, then why is Shell in Xinjiang? The developing world is one big potential energy market. Shell is one of the largest foreign investors in China, with investments totalling $2.2 billion, and plans to increase its activities to $6.8 billion by 2005. This includes a share in the $5.7-billion Nanhai petrochemicals plant, China's largest single refinery, in Guangdong. As analysts have noted, the Xinjiang pipeline, even a money-losing pipeline, offers Shell a preferential relationship with the Chinese government, in addition to a strategic position in Central Asia.

Perhaps not by accident, Shell Canada also owns 11.4 percent of the Mackenzie pipeline project and modest gas reserves in the Mackenzie Delta. It is now common for energy multinationals to hedge against their own energy plays. Petro-Canada owns gas in both the Mackenzie Valley and northern Alaska, for example, and assures its shareholders that they will benefit from either of the two competing projects.

Beijing has dubbed Xinjiang its "sea of hope" because of the region's rich energy deposits — but for many foreign multinationals, it is simply a foothold within the fastest-growing energy market in the world. "The West–East pipeline is a pathfinder for future pipelines and gas from farther afield," explained Shell Exploration (China) managing director Martin Bradshaw in 2002. "Shell has interests in Russia and Central Asia. This is a strategic reason why we're interested in the pipeline." The beginning of a huge trans-Asian pipeline network, one as complete as those seen in Europe and North America, begins in Xinjiang. "[The pipeline] will be in line with international standards and . . . will benefit all ethnic groups," claimed Bradshaw.[18]

Yet the essential formula for Central Asian strife is now solidly in place, no small thanks to North America.

Off on the other side of the Taklamakan, within a few days' drive of the Pakistani and Afghani borders, the desert outposts of Kashgar, Yarkant, and Hotan form the nominal front lines in the fight for Xinjiang. Here, Uighurs still form the majority, and the strife is even closer to the surface: photos of bloodshed are posted to buildings, if only to warn locals of police reprisals; large banners urge Uighurs to follow China's strict family-control policy, something that has seen the assassination of several Chinese enforcement officers; and used pharmaceuticals can be purchased in the local bazaar — a reminder that annual income in the Uighur heartland remains even lower than much of the rest of China. One gets the feeling that the price of natural gas and crude oil could be more than monetary. This forsaken corner of Central Asia, largely forgotten by the rest of the world, harbours

THE ONE THING THAT EVERYONE IN THE ENERGY BUSINESS CAN AGREE UPON IS THAT THE WORLD NEEDS NATURAL GAS RIGHT NOW, LOTS OF GAS.

an advanced and incomplete police state, one increasingly dedicated to the mobilization of natural resources on behalf of China and a select group of foreign multinationals.

The heart of Xinjiang, a polyglot of Uighurs, Mongols, and Kazakhs, can still be found in the family home of Rebiya, who serves up a traditional welcoming feast of rice polo, tomato and cucumber salad, savoury noodles, and mint tea. We lounge around the low table in their living room and sample the food, trading words for apricot, plum, peach, and melon. As desert winds blow through Urumqi, Rebiya's sister pulls out her dutar, a long, two-stringed guitar. After performing an old Turkic folk song, she plays a Ricky Martin tune that blazed across China during the country's first televised soccer World Cup. Over the drone of the dutar, we sing: "go, go, go, *allez, allez, allez.*" For all we know, Ricky and the dutar might be the only things holding this region together.

The one thing that everyone in the energy business can agree upon is that the world needs natural gas right now, lots of gas. In fact, as existing reserves disappear — Canada's own proven natural gas reserves fell from 97.2 trillion cubic feet (2.72 trillion cubic metres) in 1991 to 59.7 trillion cubic feet (1.67 trillion cubic metres) in 2001 — gas is proving to be one of the most strategic commodities of the twenty-first century.[19] The convergence between China and North America is striking: baseline demand for gas is growing; new demand is emerging from gas turbine and retrofitting of coal-fired power plants; and economic stagnation faces any nation that cannot provide affordable supplies of power.

What this means is that fossil fuel pipelines — and the people who control them — could soon be on a par with the railway barons of the nineteenth century. Energy suppliers hold the key to our economy, and those with transport and product will prosper. And the contrast between Canada's North and Xinjiang is striking, as only those who control their own lands have a chance at building power and dealing as equals when governments and corporations come to collect frontier resources.

Notwithstanding the negative impact of oil-sands development on gas supplies, there is a credible environmental argument in favour of Arctic gas development — and, objectionable as it may seem, even the gas of Central Asia. It has been said that a world that starves for lack of natural gas cannot, practically speaking, combat climate change or air pollution, nor could it likely develop a feasible hydrogen economy. This is, perhaps, the ultimate measure of our deep and problematic connection to global networks of resource exploration and consumption. Our energy empire is large but stretched dangerously thin. When a megaproject is touted as an environmental solution, you know that something is not quite right.

It's as though the world's energy system, left unchecked, engineers its own crisis; new solutions are often transitory, effectively mere stopgap measures within a limited economic horizon. This fault is increasingly evident outside of North America, where economic pressures and lack of wealth show systemic flaws more clearly. For example, China's incredible thirst for energy feeds not only a booming export market accelerated by North America's big-box stores but also the hungry appetites of an emergent middle class across Asia, one that rivals 1950s America for its growth in rates of car ownership and consumer accumulation. Faced with today's estimated 170 million itinerant workers and continued labour uprisings, Mao Zedong would have been hard-pressed to discern parts of modern China from the industrial England of Karl Marx.

Now, more than ever, our industrial borderlands are places where we plumb the future and track new history — whether it is a high-tech regime of repression and resource extraction, largely condoned by China's trading partners, or, closer to home, the efforts of Northern Aboriginals to build and own part of the pipeline that ships their resource birthright southward. Understanding the full cost of power is perhaps one of the biggest but least-appreciated problems of the twenty-first century. If the world's energy frontiers are any indication, our collective weakness is an inability to anticipate long-term scenarios. At some point — and we may not fully realize when this happens — North Americans could end up investing more effort and energy into shoring up dysfunction than enjoying the rich quality of life that cheap energy has provided. So the simple question still stands: What happens when the power runs out?

A HISTORY OF ENERGY (CONTINUED)

1615   Salomon de Caux illustrates the theory of the expansion and condensation of steam in his book *The Causes of Moving Forces*. For this, he will be deemed the original architect of the steam engine.

1629   The first impulse turbine, a stamping mill, using a steam-powered turbine, is invented by the Italian architect Giovanni Branca.

1687   In his publication *Principia*, the English mathematician and physicist Isaac Newton presents his "Laws of Motion," setting out the science of mechanics and forces. These fundamental ideas will become the basis for our understanding of energy.

1712   In response to his customers' problems of flooding in their mines, ironmonger Thomas Newcomen invents the first practical coal-fired steam engine to pump water out of mines. The phenomenally successful Newcomen engines will be manufactured for more than 100 years.

1719   The first written reference to Western Canada's oil sands appears in Henry Kelsey's journal. Kelsey, a fur trader for the Hudson's Bay Company, received a sample of oil-sands bitumen from the Cree warrior Wa-Pa-Sun.

1720   The first organized coal mine in North America opens at Port Morien, Cape Breton, Nova Scotia.

1752   The French scientist Thomas-Francois D'Alibard first captures electricity, using a 15-metre-long rod to draw down the "electric fluid" of lightning. Ultimately, credit for the invention of the lightning rod will go to Benjamin Franklin, however. In thirty years, there will be nearly 400 lightning rods in Philadelphia.

1765   James Watt alters the steam engine to make it faster, safer, and more fuel-efficient by condensing steam in a separate vessel. The new design will change very little over the next century and will shape the future of industry and transportation. In 1882, the unit of electrical power will be given Watt's name by the British Association.

▼ CONTINUED ON PAGE 82

03.

SHAPING AN INTEGRATED ENERGY FUTURE

LEN BOLGER, EDDY ISAACS

EDITORS' INTRODUCTION

Although the natural gas deposits in Canada's North are significant, they pale in comparison to the Alberta oil sands. Set to be a crowning jewel of the Alberta energy sector, with the potential to provide the equivalent of 2.5 trillion barrels of oil, the Alberta oil sands are one of the largest energy deposits anywhere in the world. The energy comes in the form of bitumen, a thick, tar-like substance which can be separated into oil called "synthetic crude."

Alberta has been the centre of the Canadian energy industry ever since the discovery of oil at Leduc, Alberta — just south of Edmonton — on February 13, 1947, by Imperial Oil. The find was a welcome but unexpected coup for Imperial. A discovery of crude oil in Turner Valley in 1914 had given rise to speculation about the potential of large-scale energy deposits beneath the Albertan prairies. But Imperial Oil alone had drilled 133 "dry holes," and there was every indication that Leduc would be another disappointment. Instead, it was the catalyst for an industry of enormous importance to the country.

But Alberta's incredible good fortune as the repository of much of Canada's energy reserves has too often pitted much of the Alberta energy industry against the environmental community. Of course, this polarization does no one any good, and more recently there is every sign an informed, constructive dialogue is taking place to close this energy divide. Which is essential because despite the great potential of solar, wind, and other renewable energy sources, most experts agree that a vast proportion of our energy will continue to come from fossil fuels for the near future. And so, the country will be widely reliant on the energy sector to advance the ingenuity required to provide a more sustainable energy future.

When we first went looking for someone to write a definitive article about the ingenuity coming out of the Albertan energy sector, we did not have much success, largely because we were viewed with suspicion by people in the conventional energy sector. Surely two Eastern-based editors with an interest in sustainable development would not give a fair hearing to Alberta. Fortunately, we met a man named Don Simpson. A big and amiable gentleman who is also fabulously well regarded on both sides of this energy divide, Simpson is a quintessential Canadian: warm, friendly, and passionately interested in the world.

Recently, Simpson has worked with AERI — the Alberta Energy Research Institute — developing an innovation plan for the future of the Albertan-based energy sector. He was able to introduce us to the respected heads of AERI, Len Bolger and Eddy Isaacs, who provided this chapter about the sector's efforts to develop ingenuity for our energy future.

THE CONTINUING PROSPERITY and high quality of life of Canadians depend on a secure supply of affordable energy, produced with minimal impact on the natural environment. Our ability to sustain this supply will depend on two factors. First, there must be a critical mass of people with the right knowledge and a passion for achieving this goal. Second, these people must be able to become experts at collaboration, applying their diverse skills, perspectives, and interests to the creation of technological breakthroughs.

You may be surprised to hear people like us — an engineer and a scientist with together over seventy years of experience in the Alberta energy industry — citing a "soft" factor like collaboration as the key determinant of Canada's energy future. Rest assured we're not talking vaguely about people getting along together. We're talking about a disciplined process of getting diverse thinking out on the table, seeing new connections, and co-creating innovative solutions to our most pressing problems. We're talking about structures that will allow all the players to put out their ideas and work together effectively to capitalize on opportunities, without constantly running up against organizational barriers.

The need for collaboration is not new. In this chapter we'll highlight how some unique partnerships have produced significant breakthroughs in the past and contributed to our current prosperity. But we believe we're now at an important crossroads. Our ability to use the available resources in a responsible way is limited by our current technology. To do the applied research needed and develop it into practical commercial technology will require an unprecedented level of collaboration amongst many players.

THE NEED FOR INTEGRATED, BREAKTHROUGH SOLUTIONS

We believe that transforming the way we collaborate is key for two main reasons. The first reason is that we now think the big opportunities are in integrated solutions. In the past, the strategy was to focus on one form of energy, pick a priority (like developing our ability to produce bitumen from the oil sands), and concentrate investment in that area. We've learned to appreciate, however, that the various components of the energy industry in Western Canada — conventional oil and natural gas, heavy oil, bitumen, coal, coal bed methane, petrochemicals, and renewable sources — are tightly interconnected. The solution for the petrochemical industry's need for new "feedstocks," the raw material used to produce products, may be a by-product of the oil-sands industry. The need to reduce the oil-sands industry's dependency on natural gas may be solved by coal or by waste from petroleum, forestry, and agriculture. The problem of dealing with the resulting carbon dioxide (CO_2), a greenhouse gas, may be solved by using it to get more oil out of existing fields — trapping the CO_2 safely underground at the same time — or to produce methane (the main component of natural gas) from coal beds to help offset the decline in natural gas. We'll look more closely at these integrated opportunities later. The point is that if each industry sector tries to solve its problems in isolation, much potential value will be lost.

The second reason we're advocating more collaboration is that incremental improvement is not enough. We need break-through innovations, the kind of innovations that require people with pieces of the solution to work together in new ways.

The oil sands provide a good example of the magnitude of the challenges we're facing. As early as the 1920s, a few vision-aries were pursuing the crazy idea that bitumen, a sticky sub-stance that looks like frozen molasses, could be turned into a valuable source of energy. These pioneers have driven some amazing technological advances over the past seventy to eighty years. Unfortunately we've reached the limits of what current technology can do. We're using natural gas — a lot of it — to produce the hydrogen needed to upgrade the bitumen into some-thing of greater value. Our natural gas resources are starting to decline and will become more expensive. The current recovery processes use a great deal of water, at a time when we're all becoming more concerned about water supply and quality. Emissions of greenhouse gases from the process are high. These problems will not be solved by incremental change.

Solving these challenges will require the coming together of three things. First, there must be strategic research that is clearly directed towards an end goal. Second, there must be patient capital — as opposed to investment made in anticipation of quick payoffs. And finally, there must be the know-how to turn research results into commercially viable solutions. It's because we need all three that active collaboration is key: among industry, universities and other research providers across Canada, and federal and provincial governments. We also need to be continually scouring the world for technology created elsewhere that we might learn from or adapt to our needs. We must seek opportunities to leverage inter-national sources of funding by partnering with people across and outside Canada on common issues. We're now exploring new ways of building networks that will allow us to do all this.

Before we look at what we can learn from the past and our view of the future, we want to emphasize two considerations that underlie our thinking, but that may not be universally understood and accepted. The first is that Canada and the rest of the world

ALL THE NUMBERS WE ANALYZE IN THE COURSE OF OUR WORK TELL US NOT TO EXPECT THE COMBINATION OF CONSERVATION EFFORTS AND DEVELOPMENT OF RENEWABLE SOURCES TO MEET DOMESTIC AND GLOBAL ENERGY NEEDS FOR THE NEXT FORTY TO FIFTY YEARS.

will continue to depend heavily on energy from hydrocarbons such as oil, gas, and coal for the foreseeable future. The second is that when it comes to production of this energy, "business as usual" is not an option.

THE HYDROCARBON PRESENT

We recognize that investments in renewable energy are critical to our long-term future. We also believe that there are great gains to be made in improving energy efficiency across the entire value chain, from large-scale energy production to the choices made each day by energy consumers.

At the same time, all the numbers we analyze in the course of our work tell us not to expect the combination of conservation efforts and development of renewable sources to meet domestic and global energy needs for the next forty to fifty years. The Paris-based International Energy Agency (IEA), for example, has predicted 70-percent growth in worldwide demand for primary sources of energy by 2030. The IEA expects 88 percent of this increased demand to be met by oil, natural gas, and coal. It predicts that production of nuclear energy and hydro will remain virtually flat, however. While production from other renewable sources — primarily wind — is expected to almost double, these sources will still meet only 2.5 percent of overall demand.[1] Many years of research and development will be needed to overcome the cost, technology, and infrastructure limitations to energy production from renewable sources.

In the meantime, we will continue to rely on energy from hydrocarbon sources. We in Canada have an advanced economy and a prosperous lifestyle that depend on affordable energy. We also have an abundant supply of hydrocarbon reserves. When we

CANADA HAS ONE OF THE LARGEST SUPPLIES OF HYDROCARBONS IN THE WORLD.

talk about reserves, we mean proven sources of hydrocarbons from which energy can be produced using technology available today. This doesn't include the vast potential sources of hydrocarbons — resources that experts suspect exist but can't yet prove, or that can't be extracted at a competitive cost using available technology.

Estimates of exactly how big those reserves are vary widely depending on the methods and assumptions used. One thing is clear, however: The international community's recent acceptance that energy from Alberta's oil sands is recoverable with existing technology and market conditions has changed the picture dramatically. At this year's International Energy Outlook conference, a leader from the U.S. Department of Energy cited figures from the *Oil and Gas Journal* that raised Canada's proven oil reserves to 180 billion barrels from 4.9 billion barrels, now that the oil sands are included. This makes Canada the second-largest holder of proven oil reserves in the world, behind Saudi Arabia and ahead of Iraq.[2] If you add coal and natural gas, by most estimates Canada has one of the largest supplies of hydrocarbons in the world.

Energy directly or indirectly accounts for about one-half of the Alberta economy. The Canadian Association of Petroleum Producers calculates that the oil and gas sector contributes some $15.6 billion annually to both the federal and provincial governments through royalties and income taxes alone.[3] So our focus in Alberta — our contribution to Canada's economy and to fueling the future — is to partner with other interested provinces to figure out how to get more of these resources out of the ground, get more value from the resources, and do it in a way that protects our land, air, water, health, and quality of life.

The last part of this statement is very important to us. Precisely because the production of energy from hydrocarbons comes at a significant environmental cost, the energy industry is very aware of the urgent need to reduce this impact. Back in the mid-1980s, Jack MacLeod, then president of Shell Canada, introduced the "triple bottom line" mindset that is now common in the industry. A triple bottom line mindset is one that considers the environmental and social "bottom lines" of equal importance to the financial bottom line. Leading Alberta-based companies like Suncor, Syncrude, EnCana, Nexen, TransAlta, Shell Canada, and many others have public sustainability commitments and report regularly on environmental performance.

Companies across the industry are working together to find ways of increasing energy efficiency and reducing emissions. In 1999, for example, the Canadian Association of Petroleum Producers in partnership with the Clean Air Strategic Alliance set a target to reduce flaring by 60 to 70 percent by 2007. Flaring is the practice of using controlled burning to dispose of waste gases that occur in the production of crude oil. This practice contributes to emissions of sulphur dioxide and greenhouse gases. By 2001, industry had already reduced flaring in Alberta by 53 percent.[4]

Improvements are being made, but we have a long way to go. Hence the need to invest in finding and implementing break-through technology that will allow us to realize the value of hydrocarbon resources with minimal environmental impact.

A HYDROGEN FUTURE?

We see our work as a necessary stage in the long-term transition from a hydrocarbon economy to an economy based on some form, or forms, of renewable energy. There are indications that hydrogen has good potential to play an important role in that future. Already Alberta is producing about 70 percent of the hydrogen generated in Canada, much of it for industrial use in the upgrading and refining of oil. Significant sources of excess hydrogen exist here, and the advent of fuel-cell technology to convert this hydrogen into useable electric energy would be a significant advance.

This won't be easy. Pure hydrogen doesn't exist in a natural state. You have to produce it from a hydrocarbon source such as oil or natural gas, or from water. Hydrogen produced from water through a process of electrolysis causes no greenhouse gases or other emissions, as long as the electricity used to break down water into its elements (hydrogen and oxygen) is produced from a "clean" source such as wind or nuclear. With current technology, this process uses a great deal of electricity and is very expensive. Hydrogen produced from natural gas is less expensive, but we've already noted that supplies of natural gas are starting to decline and becoming more expensive. It's most feasible in the near and mid-terms to expect to see hydrogen produced from coal and other low-value hydrocarbons. It may surprise some to learn that this process produces near-zero emissions. It's much cleaner than the process currently used to turn coal into electricity.

Hydrogen's big attraction is its potential as a means of storing energy produced by a renewable source, such as windmills, for use when the wind isn't blowing. The downside is that we'll have to invent reliable technology to make it safe. Hydrogen needs to be under very high pressure, so safe storage is difficult and expensive. There have been many successful projects demonstrating that hydrogen fuel cells can be used to run vehicles and generate power. However, much more work on storage and transport is required before fuel cells will be competitive and widespread commercial application can be achieved.

Then we'll need to create infrastructure for the delivery of hydrogen, or for the generation of hydrogen on-site when it's needed. That challenge shouldn't be underestimated. It's easy to take for granted the very complex infrastructure that allows you to fill your car with gasoline when you need it. Did you know that the gasoline you buy in the winter is different from the gas you

SO WHILE WE WORK TOWARDS CREATING A POSSIBLE HYDROGEN FUTURE, WE NEED A PARALLEL EFFORT TO MAKE BETTER, ENVIRONMENTALLY RESPONSIBLE USE OF COAL, CONVENTIONAL OIL, HEAVY OIL, BITUMEN, AND NATURAL GAS.

buy in the summer? It has to be more volatile so that it will flow and vaporize at colder temperatures. If you used the same gas in the summer, the engine would vapour-lock and stop. The gas you buy in Kamloops is different from the gas you buy in Toronto because it needs to work at a higher altitude. Providing this convenience for customers involves significant scheduling and logistical challenges. We've no idea now what challenges a hydrogen delivery system will present, but we can be reasonably sure they will exist.

So while we work towards creating a possible hydrogen future, we need a parallel effort to make better, environmentally responsible use of coal, conventional oil, heavy oil, bitumen, and natural gas.

"BUSINESS AS USUAL" IS NOT AN OPTION

Some people believe that the status quo is just fine, and wonder why we're worried about investing in the future. If that thinking had prevailed in 1974, the year the Alberta government established the Alberta Oil Sands Technology and Research Authority (AOSTRA) with an initial investment fund of $100 million, we would already be in trouble.

In 1973, 92 percent of the oil produced in Alberta was "light" oil that's under high pressure in the ground and flows naturally on its own. "Heavy" oil — oil that doesn't flow easily and

EXPERTS NOW ESTIMATE THAT CONVENTIONAL NATURAL GAS PRODUCTION HAS PEAKED AND WILL START TO DECLINE WITHIN THREE YEARS.

must be pushed out of the ground using special techniques — accounted for 4 percent of production. The remaining 4 percent came from the bitumen in the oil sands, which is essentially a solid oil. Alberta's premier at the time, Peter Lougheed, was a visionary who saw that conventional light oil production had peaked and was beginning to decline. In fact, by 2001 it had declined by 60 percent. The shortfall has been filled by increased production of heavy oil using new extraction techniques (16 percent of total oil production in 2001) and the extraction of bitumen from the oil sands (43 percent of total production in 2001). The Alberta Energy and Utilities Board projects that by 2010, 75 percent of total oil production will be from the oil sands.[5]

Experts now estimate that conventional natural gas production has peaked and will start to decline within three years. We have no shortage of coal, but coal is widely perceived to be a dirty fuel and many are arguing that we should stop using it altogether. Coal is actually much cleaner today than it was thirty years ago. The Coal Association of Canada reports that while coal-fired electrical generation capacity in North America has almost tripled over the last thirty years, related sulphur dioxide and nitrous oxides emissions have been reduced by 80 percent and 55 percent respectively.[6] However, there is still much work to be done in coal-fired energy technology to consider its use an environmentally responsible option over the long term.

THE GOOD NEWS

We do see concrete possibilities for solving these problems with the right technologies. Our current methods of extracting light oil, for example, are leaving about 70 percent of the oil in the ground. Once the pressure has dropped to the point where the oil stops flowing naturally and it's no longer feasible to push it out with water, we abandon the field. One opportunity lies in pumping CO_2 into the ground to mix with the oil, increase the pressure, and start the flow again. CO_2 is a greenhouse gas and an unwanted by-product of many industrial processes. There are good indications that pumping it into the ground, if done properly, will solve the problem of keeping it out of the atmosphere. Research is now underway to determine how to make underground sequestration of CO_2 a permanent solution. This is just one example of the many opportunities we see to get more resources out of the ground.

Other opportunities lie in adding greater value to what we do get out of the ground. The oil sands have been the focus of much of Alberta's investment over the past thirty years because while Canada is a minor player on the world stage in light and heavy oil (with 2.4 percent of the world's resources), we're *the* world player in bitumen. Canada has 58 percent of the world's bitumen resources, all in the Alberta oil sands.[7] The Alberta Energy and Utilities Board estimates that there are potentially

THIS RELIANCE ON NATURAL GAS IS A PROBLEM, BECAUSE IT'S SOON GOING TO BECOME FAR TOO COSTLY AND VALUABLE TO BE USED FOR BITUMEN UPGRADING. AS ONE OF OUR COLLEAGUES LIKES TO SAY, "USING NATURAL GAS TO MAKE SYNTHETIC CRUDE IS LIKE INVENTING A PROCESS THAT TURNS GOLD INTO LEAD."

315 billion barrels of recoverable bitumen in the oil sands. That translates to several hundred years of supply at current production rates.[8] But there are significant barriers in the way of our realizing full value from this resource.

We've already touched on the environmental issues: the need to reduce emissions and the amount of water used. Bitumen in its natural state has little value. It needs either hydrogen addition or carbon removal before it can be used for fuel. This is accomplished through an upgrading process that adds hydrogen and turns bitumen into synthetic crude oil. The cheapest source of hydrogen for upgrading is natural gas, which has more hydrogen in it than any other hydrocarbon. The "steam reforming" process for hydrogen releases CO_2 along with other emissions. We also use natural gas to create steam that we add to the bitumen at very high pressure to get the bitumen out of the ground, again generating significant CO_2 emissions.

This reliance on natural gas is a problem, because it's soon going to become far too costly and valuable to be used for bitumen upgrading. As one of our colleagues likes to say, "Using natural gas to make synthetic crude is like inventing a process that turns gold into lead." Without some major technological improvement, it's projected that the oil sands of Alberta will require one billion cubic feet (28 million cubic metres) of gas per day. To appreciate just how much that is, consider that the entire output of natural gas from new developments in the Mackenzie Delta is also expected by some estimates to be about one billion cubic feet of gas per day.

SEEING THE OPPORTUNITIES: AN INTEGRATED VIEW

This takes us back to the opportunities that arise when we start thinking about the energy industry as an integrated system, rather than each sector in its own silo. One option we have is to use petroleum coke (a by-product of bitumen upgrading) or coal instead of natural gas to produce hydrogen and steam. If the coke or the coal is burned in air (which is basically 80 percent nitrogen and 20 percent oxygen, with traces of other atmospheric gases), a mixture of CO_2 and nitrogen (with traces, again, of the other gases) is produced as a result. If we use pure oxygen rather than air to burn the coke or the coal , the by-product is then pure CO_2. That's important, because then we can capture this CO_2 without incurring the additional costs of separating it out, pump it into depleted conventional oil fields to build up the pressure and increase the flow of oil, and trap the CO_2 in the ground to reduce its emission as a greenhouse gas. Or we can pump the CO_2 into coal beds to free the methane. The reason for deadly explosions in coal mines is that there's a lot of methane in coal, and some of it will be released from the coal seam when there's a reduction in pressure. We can release the methane from coal on purpose by pumping CO_2 into it. The coal prefers the CO_2, absorbs it, and releases the methane, which can be used to partially offset the expected decline in natural gas production.

Here's another example. The petrochemical industry in Western Canada is heavily dependent on natural gas liquids such as ethane, which are by-products from natural gas extraction. These are the raw materials for petrochemical production. Petrochemicals are used in the production of plastics — everything from plastic wrap to kitchenware to siding on homes. Again, declining reserves and increasing costs create a scenario in which the competitiveness of the entire industry is threatened if an alternative is not found. What we've learned is that some of the by-products of oil-sands processing have the potential to be used as feedstocks for the petrochemical industry. We just need to build the right technology to make this happen.

A HISTORY OF INNOVATION

We have no doubt that the new technologies can be invented or adapted to meet our needs. Canadians are self-reliant and innovative. We base our confidence on a long history of successful innovation, led by entrepreneurs and visionaries. The story of the oil sands, for example, was shaped by pioneers such as Karl Clark, a scientist who worked with the Alberta Research Council (ARC) when it was established in 1921 — in itself an innovative organization as the first provincial science and technology research body in Canada. In 1929, Clark discovered and patented a hot water flotation process that is to this day the most viable method of extracting oil from the sand. Clark and his co-workers built a pilot facility near Fort McMurray before the outbreak of World War II. With the boom of conventional oil after the war, people lost interest in the gooey bitumen, but the success of today's oil-sands production companies is directly attributable to the work of people like Clark, Sidney Ellis, R. C. Fitzsimmons, and Max Bell.

It's also attributable to the leadership of visionary companies like Syncrude Canada Ltd. and Suncor Energy. Syncrude, for example, is the world's largest producer of synthetic crude oil from oil sands and the largest oil producer in Canada. Many Canadians are not aware that the development of Syncrude was a huge industrial undertaking, equivalent in scope to the building of the St. Lawrence Seaway. Hundreds of different operations are involved in producing its product — to mine the sand, extract the bitumen, and upgrade it into a synthetic crude oil called Syncrude Sweet Blend.

The CEO of Syncrude, Eric Newell, has a passionate belief in the future of the oil sands that is highly contagious. The thousands of people who have contributed to the success of Syncrude to date have been motivated to persevere in the face of many

INNOVATIONS THAT SYNCRUDE AND OTHER INDUSTRY PLAYERS LIKE SUNCOR HAVE IMPLEMENTED OVER THE YEARS THROUGH A LEARN-BY-DOING APPROACH HAVE BROUGHT THE PER-BARREL COST OF PRODUCTION DOWN BY OVER 50 PERCENT WHILE REDUCING CO_2 EMISSIONS PER BARREL BY 45 PERCENT.

setbacks, learn from failures, and refuse to be defeated. The environment of Mildred Lake, Syncrude's main field, just north of Fort McMurray, is harsh, with temperatures falling to minus 40 degrees Celsius for long periods in winter. Oil-sands mining requires much outdoor work, and in the early days mining equipment designed for warmer climates broke down continually under these extreme conditions. The teeth in the huge excavators quickly eroded in the cold, and the cost of replacing these teeth was a problem that was only overcome by finding new materials equal to the challenge. Innovations that Syncrude and other industry players like Suncor have implemented over the years through a learn-by-doing approach have brought the per-barrel cost of production down by over 50 percent while reducing CO_2 emissions per barrel by 45 percent.

Large and committed teams of people are required to operate Syncrude and Suncor, and it hasn't escaped their attention that there are many talented Aboriginal people in the community who need only some training to be able to contribute. Syncrude is now the largest Canadian industrial employer of Aboriginal people, who make up 11 percent of the total employee population of 4,000. Syncrude now supplies 13 percent of Canada's petroleum requirements, producing 81.4 million barrels in 2001, and spends more than $30 million annually on research and development. The company states on its Web site that it has returned more than $5 billion to date to the people of Alberta and the rest of Canada in the form of royalty payments and federal and provincial taxes.[9]

A TRADITION OF ENTREPRENEURSHIP

In celebrating a visionary leader like Eric Newell of Syncrude — and he's only one of many we could profile here — we're not suggesting that innovation depends on one champion at the top of an organization. We don't believe in the lone-cowboy hero who rides in and single-handedly saves the day. We could point to many examples where teams of people with no conferred "authority" got a good idea and decided to pursue it. There was the team at Imperial Oil, for example, who succeeded in proving the potential for large-scale commercial extraction of heavy oil. We said earlier that heavy oil is very viscous and does not flow naturally like light oil. The idea that perhaps you could extract heavy oil by injecting steam into the reservoir was dismissed by the experts, who concluded that the steam would not penetrate very far into the formation because the viscous oil forms a barrier. A small team nevertheless convinced their management to let them try, proved that it would work, and established the pilot from which an entire new industry in heavy oil production was born at Cold Lake. It now produces in the order of 110,000 barrels per day.

Similarly, the team at Shell Canada who promoted the idea of building a refinery that would run on synthetic crude has made a huge contribution to the financial viability of the oil sands. To understand this, it's important to know that not all crude oil is created equal. Compare a barrel of conventional light crude oil from one field in Alberta with a barrel that came from another field not far away and you'll find that the molecular composition varies quite a bit. That's a problem for the refineries that turn crude oil into gasoline and diesel and so on, because what they produce has to meet a standard set of specifications. The refineries have to configure their processing to work for a specific type of crude, and then make sure their supply stays within the acceptable ranges of variation to fit that configuration. As a result, most refineries can't add more than 10 percent synthetic crude to their supply. Shell's Edmonton refinery is currently the only refinery in the world configured to take only synthetic crude. In the beginning it bought synthetic crude from others to upgrade. Then it did a kind of "backward integration" in 2003 when it opened its own mine and started producing and shipping the bitumen itself.

In the process of learning to refine synthetic crude, an entrepreneurially minded group of Shell employees realized they could turn a problem into a profitable opportunity. The problem they were having with the composition of the synthetic crude could be solved by removing a certain type of molecule. The opportunity was that this molecule is the perfect raw material to produce styrene, used to make products such as foam cups and meat trays. Again, there were technology challenges involved that were overcome by committed teams learning by doing. They started producing styrene in 1984, much of it for export. This is a great example of the principle of "value-added production" at work.

Big corporations aren't the only drivers of innovation. The Seaman brothers, led by the brother most Albertans know as "Doc," came from Saskatchewan years ago and proved themselves to be very successful wildcatters. A wildcatter is the oil patch equivalent of a water diviner. They're independents with the instinct and guts to drill holes in exactly the right places to find oil. The good ones become very wealthy, and the Seaman brothers were good. Instead of retiring to somewhere warm to enjoy their earnings, they've chosen to stay in Alberta and invest in small, entrepreneurial companies that are developing alternative energy methods and technologies to solve many of the industry's environmental and efficiency problems.

Entrepreneurial companies like the Computer Modelling Group are driving important advances in three-dimensional modelling techniques that help companies predict how much oil they can produce from a given reservoir and determine how best to get it out of the ground. It's important to understand that "reservoirs" aren't big pools of underground oil. The oil comes from porous rock that once was an ocean bed. You might picture this rock as a less porous version of a giant sponge holding oil, natural gas, and water. Production companies need to decide where and how to drill to have the best chance of getting at the oil and gas without too much water getting into the well and making the oil production too costly. They collect raw data through a variety of means: seismic technology, analysis of rock samples, "wireline logging," which involves putting a tool down a hole on a wire to check the density of the rock, and so on. What the modelling technology does is to turn all this raw data into a three-dimensional picture of underground reservoirs that is really useful to the decision makers. This technology is now being used all over the world.

THE CRITICAL ROLE OF GOVERNMENTS

People like the Seaman brothers have an important role to play in driving the innovation needed to sustain a secure supply of affordable, responsible energy. However, they can hardly be expected to take on the challenge single-handed. Nor can industry be expected to make large investments in research and development that have no hope of return for many years. Some of the most important advances with the potential for big payoffs can take twenty years and several hundred million dollars. When industry ceases to be profitable, it ceases to exist. There's a clear role, therefore, for governments to provide leadership and investment, as the stewards of these resources on behalf of all Canadians.

We've used the plural "governments" because we mean federal and multiple provincial governments, not just the Government of Alberta. As members of the leadership team of the Alberta Energy Research Institute (AERI), the successor to AOSTRA, we're very aware that the issues we're talking about transcend provincial borders. The resources are spread through the Western Canada Sedimentary Basin, which takes in large parts of British Columbia and Saskatchewan and a bit of Manitoba, and so we're really talking about an integrated Western Canadian energy industry. Also, we need to apply the best brains — wherever we find them — to the technology challenges we're facing. And finally, the benefits of this investment will accrue to all Canadians.

The most obvious benefit is increased confidence that there will be a secure and dependable supply of affordable, responsible energy to fuel the needs of Canadians now and in future generations. Less well understood is how many of the direct benefits of new energy investment accrue to companies outside of Western Canada. In a typical $5-billion oil-sands development project, more than 60 percent of the capital is spent outside Alberta, on goods manufactured in other Canadian provinces.[10] Taxes flow back to the federal government, where they can be applied to important social priorities.

LOCAL KNOWLEDGE IS KEY: THE "HUFF AND PUFF" STORY

It's rarely possible to simply purchase a technology developed elsewhere and apply it in the Canadian energy industry. Some years ago there was a lot of interest in technology that was being used in California to produce heavy oil, including the cyclic steam injection system called "huff and puff." Ninety percent of the heavy reserves in the oil sands can't be excavated with mining equipment because they lie too deep underground, and it was thought that "huff and puff" might be the solution. It wasn't. But the many years spent trying and the dollars invested produced a lot of learning that eventually resulted in a made-in-Canada solution: steam-assisted gravity drainage (SAGD).

The solution became possible only after the invention of horizontal drilling and improvements in three-dimensional seismic technology. The seismic technology allows you to use sound waves to see what's under the surface without excavating — not unlike ultrasound. With this knowledge, you can make better decisions about where to place wells. The SAGD process involves drilling two horizontal wells, injecting steam from one to "melt" the heavy oil and bitumen and make it fluid, then draining the oil by gravity from that chamber into the second well, which is used for production.

According to independent estimates, the invention of SAGD made enough additional reserves recoverable to support 340 commercial projects, each producing 25,000 barrels per day for sixty years. Across Canada, each project would create 823 permanent jobs, of which 540 would be in Alberta. There's also good news on the environmental front. Traditional oil-sands mining involves disturbing large tracts of land and then reclaiming them once they're mined. The SAGD process requires only two small holes. While we don't pretend this has no environmental impact, it's much less disruptive to ecosystems. Compared to other thermal recovery processes, the new process has 35 percent less greenhouse gas emissions.[11]

The story of SAGD is a perfect example of the kind of industry–government collaboration we believe is necessary. The first steps in developing the concept were made by industry, which quickly realized that the technical and financial uncertainties were greater than it could take on. AOSTRA saw too much good potential in the technology for it to be abandoned altogether. It stepped up to the plate and invested over $80 million of public money over fifteen years to field test the technology and prove its technical and economic viability. Industry became heavily involved again at the pilot stage and invested $900 million in their own field pilots, turning the beta-test into a commercially viable technology.

Building on this knowledge, a hybrid steam-solvent process has recently been invented. Preliminary findings indicate that this process can increase oil production by 19 percent over SAGD with half the water use.[12]

THE PUBLIC–PRIVATE EXPERIMENT

The role that public–private co-operation has played in the history of the energy industry in Alberta to date is, to our knowledge, unique. The Government of Alberta relies on the Alberta Science and Research Authority (ASRA) in making decisions on research investment. When we explain to colleagues elsewhere in the world that ASRA's board is made up mostly of private-sector members, we're usually met with some very surprised looks. It seems logical to us, however, that industry leaders would be heavily involved in making the strategic decisions on where money for research and technology development should be invested, since we're counting on those industry leaders to apply the resulting technology to benefit all of us.

Our organization, AERI, is part of ASRA, and has been given a broad mandate to look at all forms of energy in Alberta. AERI is also governed by a board that includes leaders from business and industry. We're charged with deciding how to invest funding earmarked by the Alberta government for energy research, and how to manage that investment. As we consider the complex challenges that face us at this crossroads, we see an urgent need to take to a new level the collaborative approach that has worked well for Alberta in the past.

Innovation is a complex process. It requires research, but it also needs investors who believe they can profit by commercializing the research, people capable of implementing the innovation, policy-makers to ensure that the innovation is safe and in the public interest, and a receptive market of people ready to adopt the innovation.

Academic institutions and other research providers such as the Alberta Research Council, the federal Natural Resources Canada, and the National Research Council laboratories need to be involved from the beginning in the process of shaping and launching the innovation agenda. Universities and colleges have a critical role to play in building the all-important human capacity we're expecting to sustain the hydrocarbon economy in the near term and build the integrated energy economy of the future. We need to be thinking now about what combination of technical know-how and attitudes — towards environmental responsibility, towards working as part of a collaborative team — will best equip today's students to play their role.

These institutions are also home to the scientists and other experts who will need to do the basic research and invent the science that will make the new technology possible. Typically, all this talent will be engaged in finding answers to problems that are inherently interesting but may have no application to the pressing problems threatening our future prosperity. This basic research is essential for increasing our store of scientific knowledge. It has served us well in the past and must continue. Increasingly, however, we need some of this talent to take on what we call "mission-oriented" research focused on solving specific problems or creating a significant opportunity.

In the past, we've worried about the existence of two solitudes: academic researchers and the potential users of research. Too much brilliant research results in a published paper that's remote from the realities facing industry and doesn't get used. The time and resources required to turn an innovative idea into a viable solution can be considerable. Once you get into the "big laboratory" of the commercial market, you inevitably run up against issues you never thought of or didn't see when you were working at a smaller scale. We believe that as an empowered government body, AERI can play a role in trying to bridge this divide, providing leverage funding and structuring important initiatives so that it's in the interest of all the key players to collaborate. We've found champions in universities and in industry who share this mission and are taking the lead in their organizations to make this happen.

LEARNING TO COLLABORATE

Of course, the two-solitudes issue can hamper collaboration within a sector or an organization as easily as it does between organizations. Government ministries and departments aren't always great at working together effectively. Companies primarily working in coal haven't had much reason to talk to companies engaged in oil-sands mining and processing. Research and marketing departments within the same company have occasionally been known to be working at odds. Rivalries may possibly develop between different groups of researchers at the same institution.

Because we believe active collaboration to be the key determinant of future success for the energy industry in Canada, and because we know it doesn't always happen naturally, we've been making a conscious effort to learn how to do it better. For the past six months, AERI has been using a disciplined process called the Challenge Dialogue System™ to get key players engaged in a structured conversation designed to produce action — not just talk. This is part of our effort to move from a model that has us identifying and funding a series of discrete projects towards a more program-like approach. That means identifying a desired future state in which our priority problems are solved, assessing our current situation, and then building a portfolio of linked and complementary initiatives that need to all come together to create that future state. The future-state scenarios might be something like "A cost-competitive and sustainable supply of coal-based electricity, with all emissions (including carbon dioxide) from power plants reduced to near-zero levels." Or "More upgrading of bitumen in Alberta to ensure that more of the value of the resource remains in Canada." Or "An orderly transition from a hydrocarbon economy to a hydrogen economy based on environmentally responsible renewable technologies."

We argued earlier that these challenges are much bigger than Alberta. We've been encouraging interested partners across Canada and even outside Canada to join in a network of people actively engaged in charting future directions and participating where they can make a contribution. We're not presumptuous enough to pretend to be setting the energy innovation agenda for Canada. We just realize that while we attack the immediate issues we see and are uniquely placed to address, there will be points of connection and opportunities to collaborate with others addressing other important energy issues.

Action is starting on many fronts even as we finalize details on what this network will be called, how we will engage other important voices in the ongoing dialogue (especially groups focused on relevant social and environmental issues), and how exactly the network will operate. We've already benefited enormously from focused interaction with people from the federal and other provincial governments and key players in the energy industry in Canada.

If we all stay focused and committed, perhaps ten years from now someone will be sitting down to write the story of how a small group of people with some passion around a shared vision learned from the successes of the past, figured out how to collaborate effectively to shape an integrated energy future, and had a major impact on the social, environmental, and economic prosperity of Canadians.

A HISTORY OF ENERGY (CONTINUED)

1770   Nicholas Joseph Cugnot, a French military engineer, invents the world's first automobile, which he dubs a "*voiture en petit*," though it looks more like a steam-powered tricycle. Although it travels at only 4 km/h, its boiler needs to be refilled every fifteen minutes.

1776   Henry Cavendish first ascertains that hydrogen is a distinct element, describing it as "inflammable air from metals."

1783   Watched by Louis XVI and Marie Antoinette, the French papermakers Joseph and Etienne Montgolfier launch a sheep, a duck, and a rooster in their paper balloon filled with heated air.

1792   The Scotsman William Murdock runs a metal tube from his backyard to his living room and achieves a gas flame inside the room using coal gas. Soon after, most large factories in Britain will be using gas lighting, and eventually the system will expand to street lighting.

1800   The Italian philosopher and physicist Alessandro Volta invents the electric battery's first predecessor, the so-called voltaic pile, which produces a steady electric current. For Volta's efforts, Napoleon will make him a count in 1810, and in 1881 the electrical unit known as the volt will be named in his honour.

1802   The engineer Richard Trevithick, also known as the "Cornish giant," perfects the high-pressure steam engine in a life-sized steam road locomotive which he calls the *Puffing Devil*.

1804   Trevithick's *Penydarren* is the first steam locomotive in England, and will be the model for all future steam locomotives.

1807   For the first time, street lamps in London, England, are illuminated using coal gas.

1831   Michael Faraday, an English chemist and physicist, discovers electromagnetic induction. He will also discover the battery (electropotentials), the electric arc (plasmas), and the Faraday cage (electrostatics), although not all in the same year. All of these will lead to the development of electric motors, generators, and transformers.

▼   CONTINUED ON PAGE 104

04.

THE DAWN OF THE HYDROGEN ECONOMY

JEREMY RIFKIN

EDITORS' INTRODUCTION

During the American Civil War, five soldiers and a dog escaped captivity in a Confederate prison by sailing off in a hot-air balloon. After floating for five days and thousands of miles, they eventually landed on an abandoned island far from civilization. As the months passed, the soldiers talked often of life back home, and one of their conversations would have been of particular interest to the oil and gas companies of today. What will happen to the country, one of the soldiers asked, when the supply of coal runs out? The engineer of the group, a thin, moustached man of forty-five named Cyrus Harding, predicted that a new fuel would replace coal: water. "My friends," Harding explained, "I believe that water will one day be employed as fuel, that hydrogen and oxygen which constitute it, used singly or together, will furnish an inexhaustible source of heat and light, of an intensity of which coal is not capable."

This was an extraordinarily prescient remark, but Harding can't actually take credit for it. The man who can is the French writer Jules Verne. Harding and his four fellow Union soldiers are merely characters in Verne's 1874 survivalist novel, *The Mysterious Island* (trans. W. H. G. Kingston, The Limited Editions Club at the Garamond Press, 1959). To read the book today is to realize how closely Verne's energy prediction came to the vision of most present-day energy and automobile companies. Listen to what Robert Purcell, the executive director of General Motors, told the annual meeting of the National Petrochemical and Refiners Association in May 2000. "Our long-term vision," Purcell said, "is of a hydrogen economy." Verne's dream is no longer science fiction.

Few people have been more active in making Verne's prediction come true than Jeremy Rifkin. The author of the best-selling book *The Hydrogen Economy*, Rifkin is a leading advocate of the hydrogen economy. His vision of our energy future may not be opposed to the long-term view of the conventional oil and gas industry, but his short-term proposals certainly are. Which is why, in considering how to fuel our future, we have to look at this side of the energy spectrum as well.

This is how Jeremy Rifkin believes we ought to move onto a Jules Verne–inspired island, where everything is powered by hydrogen.

IMAGINING A WORLD WITHOUT OIL

Imagine, for a moment, a world where fossil fuels are no longer burned to generate power, heat, or light. A world no longer threatened by global warming or geopolitical conflict in the Middle East. A world where every person on Earth has access to electricity. That world now looms on the horizon.

We are in the early stages of an historic change in the way we organize the Earth's energy. The Industrial Age, which began with the carrying of coal from Newcastle several hundred years ago, is now winding down in the oil fields of the Middle East. Meanwhile, a wholly new energy regime is being readied. Hydrogen — the lightest and most abundant element in the universe — is the key to the next great energy revolution. Scientists call hydrogen the "forever fuel" because it never runs out. And when it's used to produce power, the only by-products are pure water and heat.

It's difficult to comprehend a world beyond oil when so much of the Modern Age has been fueled by the burial grounds of the Jurassic Era. We heat our homes and businesses, run our factories, power our transportation, and light our cities with fossil fuels. We grow our food and construct our buildings with materials made from fossil fuels, treat illnesses with pharmaceuticals made from fossil fuels, and produce our clothes and home appliances with petrochemicals. Virtually every aspect of modern life is made from, or powered by, fossil fuels.

Until recently, experts had been saying that we had another forty or so years of cheap, available crude oil left. Now, however, some of the world's leading petroleum geologists are suggesting that global oil production could peak and begin a steep decline much sooner, as early as 2020, at which point — assuming we don't curb our global appetite for oil — prices will go through the roof. Oil-producing countries outside of the Organization of Petroleum Exporting Countries (OPEC) are already nearing their peak production, leaving most of the remaining reserves in the politically charged Middle East. Increasing tensions between the Islamic world and the West are likely to compound the threat to our access to affordable oil. Rising oil prices will assuredly plunge developing countries even further into debt, ensuring that much of the Third World remains in the throes of poverty for years to come. In desperation, many nations may turn to dirtier fossil fuels — coal, oil from tar sands, and heavy oil — which will only worsen global warming and imperil the Earth's already beleaguered ecosystems. Looming oil shortages will make industrial life vulnerable to massive disruptions and possibly even collapse.

Hydrogen has the potential to end the world's reliance on oil from the Persian Gulf, the most politically volatile region on the planet. Indeed, making the transition to hydrogen is the best insurance against the prospects of future oil wars in the Middle East. Hydrogen will also dramatically cut down on carbon dioxide emissions and mitigate the effects of global warming. And because sources of hydrogen are so plentiful, people who have never before had access to electricity will be able to generate it.

In October 2002, the European Union became the world's first superpower to announce a long-term plan to make the transition out of fossil fuel dependency and into a renewable-based hydrogen economy. Romano Prodi, the president of the European Commission, said at the time that weaning Europe off Middle East oil and making the shift to a hydrogen future would be the next great integrating task for Europe after the introduction of the Euro, and he compared the ambitious effort to the American space program in the 1960s to put a man on the moon.

HOW HYDROGEN POWER WORKS

Hydrogen is found everywhere on Earth, yet rarely exists in nature as a free-floating gas. Instead, it has to be extracted from either hydrocarbons or water. Today, the most cost-effective way to produce commercial hydrogen is to harvest it from natural gas via a "steam reforming" process. However, greenhouse gas emissions are produced as a by-product of steam reforming. Moreover, the supply of natural gas is finite just like our oil supply, and therefore is not a dependable long-term source. There is, however, another way to produce hydrogen, one that uses no fossil fuels. Renewable sources of energy — photovoltaic cells (solar), wind, hydro, geothermal, and biomass — are increasingly being used to produce electricity. That electricity, in turn, can be used in a process called electrolysis to split water into hydrogen and oxygen. Once produced, the hydrogen can be stored and used, when needed, to generate electricity.

Storage is the key to making renewable energy economically viable. That's because when renewable energy is harnessed to produce electricity, the electricity flows immediately. But if the sun isn't shining or the wind isn't blowing or the water isn't flowing, electricity isn't generated and economic activity relying on that electricity grinds to a halt. Which is why it is important to store electricity in the form of hydrogen. To do that, some of the electricity generated from solar or wind must be used for extracting hydrogen from water. That way, society will have a continuous supply of power.

While the costs of harnessing renewable technologies and extracting hydrogen are still high, new technological break-throughs and economies of scale are dramatically reducing these costs every year. Moreover, hydrogen-powered fuel cells are two and one-half times more efficient than internal combustion engines. Meanwhile, the direct and indirect costs of oil and gas on world markets will continue to rise. As we approach the nexus between the falling price of renewables and hydrogen and the rising price of fossil fuels, the old energy regime will steadily give way to the new energy era.

THE FUTURE IS NOW

Stationary commercial fuel cells powered by hydrogen are just now being introduced for home, office, and industrial use. Portable fuel-cell cartridges will be on the market in a few years. Consumers will be able to power up their cell phones, laptops, computers, and other appliances for forty days or more with a single cartridge. The major automakers have already spent over US$2 billion developing hydrogen cars, buses, and trucks, and the first mass-produced vehicles are expected to be on the road beginning in 2009.

The hydrogen economy will make possible a vast change in the way power is distributed, with far-reaching consequences for society. Today's centralized, top-down flow of energy, controlled by global oil companies and utilities, could become obsolete. In the new era, every human being with access to renewable energy sources could become a *producer* as well as a consumer of his or her own energy, using so-called "distributed generation." When millions of end-users connect their fuel cells into local, regional, and national hydrogen energy webs (HEWs), using the same design principles and smart technologies that made possible the World Wide Web, they can begin to share energy — peer to peer — creating a new, decentralized form of energy generation and use.

In the new hydrogen fuel-cell era, even the automobile itself will become a "power station on wheels," with a generating capacity of twenty kilowatts. The average house requires only two

THE TRULY GREAT ECONOMIC REVOLUTIONS IN HISTORY OCCUR WHEN NEW COMMUNICATION TECHNOLOGIES FUSE WITH NEW ENERGY REGIMES TO CREATE WHOLLY NEW ECONOMIC PARADIGMS.

to four kilowatts of power. Since cars are parked most of the time, owners will be able to plug them into the home, the office, or the main interactive electricity network, during non-use hours, selling the electricity they produce back to the grid. If just 25 percent of drivers used their vehicles as mini–power plants, we could eliminate all the giant, environmentally polluting power plants on which we now depend.

THE MARRIAGE OF SOFTWARE, COMMUNICATIONS, AND HYDROGEN

The truly great economic revolutions in history occur when new communication technologies fuse with new energy regimes to create wholly new economic paradigms. The introduction of the printing press in the 1400s, for example, established a new form of communication that, when later combined with coal and steam technology, gave birth to the industrial revolution. Print provided a form of communication that was information intensive and quick enough to co-ordinate a world propelled by steam power. It would not have been possible to co-ordinate the increase in speed, pace, flow, density, and interactivity of commercial and social life — made possible by steam power — by relying on script or oral communication technologies. Similarly, the telegraph and, later, the telephone provided forms of communication that were fast enough to accommodate the quickened pace, flow, density, and interactivity made possible when coal gave way to an even more agile hydrocarbon, crude oil.

Today, hydrogen and the new fuel-cell, distributed-generation technology are beginning to fuse with the computer and communications revolution to create a wholly new economic era. Before the distributed generation of hydrogen can be fully achieved, however, changes in the existing power grid will have to be made. That's where the software and communications revolution fits in. Connecting thousands and then millions of fuel cells to main grids will require sophisticated dispatch and control mechanisms to route energy traffic during peak and non-peak periods. A U.S. company called Encorp has already developed a software program for remote monitoring and control that would automatically switch local generators onto the main grid during peak loads, when more auxiliary energy was required. Retrofitted existing systems are estimated to run at about US$100 per kilowatt-hour, which is still less costly than building new capacity.

An inherent problem with the existing power grid is that it was designed to ensure a one-way flow of energy from a central source to all of the end-users. It is no wonder that Kurt Yeager, president of the Electric Power Research Institute (EPRI), recently remarked that "the current power infrastructure is as incompatible with the future as horse trails were to automobiles." In many ways, the current grid resembles the state of the broadcast industry before the advent of the World Wide Web, when connections flowed in only one direction, from the media source to the viewing audience. Today's transmission systems are not set up to direct specific quantities of energy to specific parts of the grid. The result is that power flows all over the place, often causing congestion and energy loss. A new technology developed by EPRI called FACTS — short for Flexible Alternative Current Transmission System — gives transmission companies the capacity to deliver measured quantities of power to specified areas of the grid.

The integration of state-of-the-art computer hardware and software transforms the centralized grid into a fully interactive intelligent energy network. Sensors and intelligent agents embedded throughout the system can provide up-to-the-minute information on energy conditions, allowing current to flow exactly where and when it is needed and at the cheapest price. For example, Sage Systems, in the United States, has created a software program that allows utilities to "shed load instantly" if the system is at peak and stressed to the limit, by "setting back a few thousand customers' thermostats by 2 degrees . . . [with] a single command over the Internet." Another new product, Aladyn, allows users, with the aid of a Web browser, to monitor and make changes in the energy used by home appliances, lights, and air conditioning.

In the very near future, sensors attached to every appliance or machine powered by electricity — refrigerators, air conditioners, washing machines, security alarms — will provide up-to-the-minute information on energy prices, as well as on temperature, light, and other environmental conditions, so that factories, offices, homes, neighbourhoods, and whole communities can continuously and automatically adjust their energy requirements to one another's needs and to the energy load flowing through the system.

THE PEOPLE'S ENERGY

Whether hydrogen becomes "the people's energy" depends to a large extent on how it is harnessed in the early stages of development. The first thing to keep in mind is that with distributed generation, every family, business, neighbourhood, and community in the world will be potentially both a producer and a consumer of its own hydrogen and electricity. Because fuel cells will be located at the sites where the hydrogen and electricity are going to be produced and partially consumed, with the surplus hydrogen sold as fuel and the surplus electricity sent back onto the energy network, the ability to aggregate large numbers of producer-users into associations will be critical to energy empowerment and the advancing of the vision of democratic energy.

Like with the struggle to control the World Wide Web, we are likely to see a hard-fought and protracted battle for control over HEWs. End-users have long argued that information ought to run free over the Web. Time Warner, Microsoft, and other global software and content companies, on the other hand, have fought hard to control access to the portals of cyberspace. Expect global energy companies like BP and Royal Dutch/Shell and the world's leading power and utility companies to attempt to exercise similar control over almost every aspect of the emerging HEWs.

The aggregation of distributed generation has much in common with the aggregation of labour in the early union movement at the beginning of the twentieth century. Industrial workers, individually, were too weak to negotiate the terms of their labour contracts with management. Only by organizing collectively as a block within factories, offices, and whole industries could labour amass enough power to bargain with management. The ability to withhold labour collectively by using "the strike" gave labour a powerful tool in its campaign to shorten workweeks, improve the conditions of work, and increase both pay and benefits.

Similarly, empowering individuals and democratizing energy will require that public institutions and non-profit organizations — local governments, co-operatives, community development corporations, credit unions, and the like — jump in at the beginning of the new energy revolution and help to establish distributed-generation associations in every country.

Eventually, the end-users' combined generating power via the energy webs will exceed the power generated by the utility companies at their central plants. When that happens, it will constitute a revolution in the way energy is produced and distributed. Once the customer, the end-user, becomes the main producer and supplier of energy, power companies around the world will be forced to redefine their role if they are to survive. A few power companies are already beginning to explore new roles as bundlers of energy services and co-ordinators of energy activity on the energy web that is forming. In the new scheme of things, power companies would become "virtual utilities," assisting end-users by connecting them with one another and helping them to share their energy surplus profitably and efficiently. Co-ordinating content rather than producing it will become the mantra of power companies in the era of distributed generation.

Utility companies, interestingly enough, serve to gain — at least in the short term — from distributed generation, although, until recently, many have fought the development. Because distributed generation is targeted to the very specific energy requirements of the end-user, it is a less costly and more efficient way of providing additional power than relying on a centralized power source. It costs an American company between US$365 and US$1,100 per kilowatt-hour to install a six-mile (ten-kilometre) power line to a three-megawatt customer. A distributed-generation system can meet the same electricity requirements at a cost of between US$400 and US$500 per kilowatt-hour. Generating the electricity at or near the end-user's location also reduces the amount of energy used, because between 5 and 8 percent of the energy transported over long-distance lines is lost in the transmission.

U.S. power companies are anxious to avoid making large financial investments in capital expansion because under the new utility restructuring laws they can no longer pass the costs of new capacity investment on to their customers. And because the field is now very competitive, power companies are reluctant to take funds from their reserves to finance new capacities. The result is that they put stress on existing plants beyond their ability to keep up with demand, leading to more frequent breakdowns and power outages. The major East Coast blackout of

CHAMPIONING A FIFTY-YEAR PLAN TO BUILD A HYDROGEN ECONOMY IS A GRAND ECONOMIC VISION ON THE SCALE OF THE FIRST AND SECOND INDUSTRIAL REVOLUTIONS IN NORTH AMERICA.

August 14, 2003, is the most illuminating recent example of this. That is why a number of power companies are looking to distributed generation as a way to meet the growing commercial and consumer demand for electricity while limiting their financial exposure.

TOWARDS A THIRD INDUSTRIAL REVOLUTION

The harnessing of hydrogen will alter our way of life as fundamentally as did the introduction of coal and steam power in the nineteenth century and the shift to oil and the internal combustion engine in the twentieth century. Championing a fifty-year plan to build a hydrogen economy is a grand economic vision on the scale of the first and second industrial revolutions in North America. By taking a commanding lead in building a hydrogen infrastructure for Canada and by developing renewable resources and hydrogen technologies and related products and services, the

Canadian economy can help to set the twenty-first-century economic agenda for the rest of the world.

Investing in a hydrogen economy will reinvigorate capital markets, spur productivity, create new export markets, and increase the GDP of Canada. According to a recent study by Price Waterhouse Coopers, the hydrogen economy could generate US$1.7 trillion in new business by the year 2020. It should be emphasized that no other single economic development will have as great an effect on the global economy over the course of the next several decades.

The transition from a fossil fuel energy regime to the hydrogen age will require a dynamic partnership between Canadian business and local, provincial, and national governments. Business will provide the entrepreneurial know-how to create the software and hardware of the new hydrogen era and redesign and manage the decentralized hydrogen energy grid. Government, at every level, will need to ease the transition by partnering with business. In the early stages of creating a new energy regime and laying out a new energy infrastructure, government assistance, in the form of research and development funds, tax credits and incentives, early technology adoption agreements by government agencies and their contractors, and favourable regulatory changes will all be critical to making the transition workable.

A government–business partnership will quicken the pace of the change by helping industry to underwrite the large direct and indirect costs involved in getting to the kind of economies of scale and speed that will make the new technology and infrastructure commercially viable. All earlier energy revolutions were similarly underwritten by the forging of government–business partnerships. Canada has traditionally supported public–private partnerships and, therefore, is ideally suited to steward a new working relationship between government and industry to make the transition to the hydrogen era.

Organized labour will also benefit from the shift to the hydrogen economy. While new "smart" technologies are moving the global economy away from mass labour and towards small, professional workforces over the long term, in the short term — the next thirty years — millions of new jobs will be required to install renewable technologies in every community, reconfigure the nation's power grids, and create a hydrogen energy infrastructure. Qualitative leaps in employment always occur during periods in history when new energy regimes are being established and the accompanying infrastructures are being laid out. The harnessing of coal and steam power and the laying down of continental rail infrastructures between the end of the American civil war and the beginning of World War I created millions of jobs in North America, as did the harnessing of oil, the introduction of the internal combustion engine, the laying down of roads, and the electrification of factories and communities in the first sixty years of the twentieth century. Once operational, these new energy regimes — the first and second industrial revolutions — spawned great leaps in productivity and made possible new kinds of goods, services, and markets, resulting in the creation of still more jobs.

The key point is that fundamental changes in energy regimes and accompanying infrastructures are traditionally the source of new employment opportunities. And because the installation of renewable technologies and the establishment of a hydrogen infrastructure as well as the reconfiguration and decentralization of the nation's power grid are geographically tied, the employment generated will all be within Canada. If both the technologies and technical know-how that comprise the hydrogen economy are also produced by research institutes and Canadian-based companies, additional domestic employment will be generated.

Making the transition to the hydrogen era provides a unifying vision for the environmental movement and offers the first real hope of creating a truly sustainable global economy for future generations. By eliminating carbon dioxide altogether from the economic equation, the hydrogen economy leaps ahead of the current paltry and piecemeal efforts to reduce greenhouse gas emissions. The shift to hydrogen is a bold plan to confront, head-on, global warming, the single most dangerous problem facing humanity and the Earth in the coming century. The ambitious and uncompromising nature of the plan will reinvigorate the green spirit, especially among the young, who will likely identify with both the high-tech and the democratic vision of peer-to-peer energy sharing.

The hydrogen economy will also improve the lot of Canada's most disadvantaged citizens. The rising price of oil and gas in the years ahead will fall disproportionately on the poor. Already, the poor, many of whom have lost part-time minimum-wage jobs in the recent downturn of the global economy, are increasingly unable to pay their electricity, gas, and heating bills and cannot afford the rising price of gasoline at the pump. A Canada-wide hydrogen game plan that emphasizes the installation of renewable technologies and a hydrogen fuel-cell infrastructure in poor urban and rural communities can help to create energy independence among Canada's most vulnerable populations.

EMPOWERING THE DEVELOPING WORLD

Incredibly, 65 percent of the human population has never made a telephone call, and a third of the human race has no access to electricity. Today, the per-capita use of energy throughout the developing world is a mere one-fifteenth of the consumption enjoyed in the United States. Narrowing the gap between the haves and have-nots means first narrowing the gap between the connected and the unconnected. Lack of access to electricity is a key factor in perpetuating poverty around the world. Conversely, access to energy means more economic opportunity.

Electricity frees human labour from day-to-day survival tasks. It provides power to run farm equipment, operate small factories and craft shops, and light homes, schools, and businesses. Making the shift to a hydrogen energy regime, using renewable resources and technologies to extract the hydrogen, and creating distributed-generation energy webs that can connect communities all over the world, together hold great promise for helping to lift billions of people out of poverty.

As the price of fuel cells and accompanying appliances continues to plummet with new innovations and economies of scale, these products will become far more broadly available, just as was the case with transistor radios, computers, and cellular phones. The goal ought to be to provide stationary fuel cells for every neighbourhood and village in the developing world. Villages can install renewable energy technologies — photovoltaic, wind, biomass, etc. — to produce their own electricity and then use it to separate hydrogen from water and store the hydrogen for subsequent use in fuel cells. In rural areas where expensive commercial power lines have not yet been extended, stand-alone fuel cells can provide energy quickly and cheaply. After enough fuel cells have been leased or purchased and installed, mini–energy grids can connect urban neighbourhoods as well as rural villages into expanding energy networks. The HEW can be built organically and spread as the distributed generation becomes more widely used. The larger hydrogen fuel cells have the additional advantage of producing pure drinking water as a by-product, a not-insignificant consideration in village communities around the world where access to clean water is often a critical concern.

Distributed-generation associations need to be established throughout the developing world. Civil-society organizations, co-operatives (where they exist), micro-credit lending institutions, and local governments ought to view distributed-generation energy webs as the core strategy for building sustainable, self-sufficient communities. Breaking the cycle of dependency and despair, becoming truly "empowered," starts with access to and control over energy.

National governments and world lending institutions need to be lobbied or pressured to help provide both financial and logistical support for the creation of a hydrogen energy infrastructure. Equally important, new laws will need to be enacted to make it easier to adopt distributed generation. Public and private companies will need to be required to guarantee that distributed-generation operators have access to the main power grid and the right to sell energy back or trade it for other services.

The fossil fuel era brought with it a highly centralized energy infrastructure, and an accompanying economic infrastructure, that favoured the few over the many. Now, on the cusp of the hydrogen age, it is possible to imagine a decentralized energy

IF ALL INDIVIDUALS AND COMMUNITIES IN THE WORLD WERE TO BECOME THE PRODUCERS OF THEIR OWN ENERGY, THE RESULT WOULD BE A DRAMATIC SHIFT IN THE CONFIGURATION OF POWER.

infrastructure, the kind that could support a democratization of energy, enabling individuals, communities, and countries to claim their independence.

In the early 1990s, at the dawn of the Internet era, the demand for "universal access" to information and communications became the rallying cry for a generation of activists, consumers, citizens, and public leaders. Today, as we begin our journey into the hydrogen era, the demand for universal access to energy ought to inspire a new generation of activists to help lay the groundwork for establishing sustainable communities.

If all individuals and communities in the world were to

become the producers of their own energy, the result would be a dramatic shift in the configuration of power: no longer from the top down but from the bottom up. Local peoples would be less subject to the will of far-off centres of power. Communities would be able to produce many of their own goods and services and consume the fruits of their own labour. But, because they would also be connected via the worldwide communications webs, they would be able to share their unique commercial skills, products, and services with other communities around the planet. This kind of economic self-sufficiency becomes the starting point for global commercial interdependence, and is a far different economic reality than that in colonial regimes of the past, in which local peoples were made subservient to and dependent on powerful forces from the outside.

By redistributing power broadly to everyone, we can establish the conditions for a truly equitable sharing of the Earth's bounty. This is the essence of the politics of reglobalization from the bottom up.

TWO APPROACHES TO A HYDROGEN FUTURE

While the European Union understands that in the immediate future much of the hydrogen will have to be extracted from fossil fuels, its long-term game plan is to rely increasingly on renewable sources of energy to extract hydrogen. (The E.U. has set a target to generate 22 percent of its electricity and 12 percent of all of its energy from renewable sources of energy by 2010.)

Now, the U.S. business community is worried that it might fall behind Europe in reaching a hydrogen future and has begun to put pressure on the Bush administration to spearhead a similar effort. Even though the American president embraced the hydrogen future in his 2003 State of the Union address, in reality the energy bill he sent to Congress for deliberation focuses almost entirely on subsidizing research and development aimed at extracting hydrogen from fossil fuels and by harnessing nuclear power to the task, with little emphasis on developing renewable sources of energy to extract hydrogen. In other words, the White House would like to head into a hydrogen future without ever leaving an old-fashioned fossil fuel and nuclear energy regime. Its failure to imagine a new energy era and to take the steps to get there could put the United States significantly behind Europe as a world power by mid-century.

LAYING THE GROUNDWORK FOR THE HYDROGEN ECONOMY IN CANADA

In order to jump-start the hydrogen economy in Canada, the federal government should consider adopting a number of high-visibility initiatives. First, create a high-level working group to draft a blueprint for Canada to become an integrated hydrogen economy by the year 2050. Second, assemble a consortium of universities, technical institutes, and government laboratories to help facilitate research and development of hydrogen technology and related products and services. Third, create a working group of software, chemical, automotive, energy, and power companies to co-ordinate joint efforts to produce and market hydrogen technologies. Fourth, prepare a fiscal plan for using government funds to stimulate research and development programs — in partnership with the private sector — and to provide tax incentives for companies and consumers to produce and use hydrogen-based technologies. Fifth, establish a working group of financial institutions to work with the federal government to jointly underwrite new venture opportunities. Sixth, establish several strategically located hydrogen technology parks and provide tax credits and incentives to lure businesses from around the world that are engaged in hydrogen technology products and services to set up shop in Canada. These hydrogen technology parks can create a "synergy effect" and spur quicker development, much like Silicon Valley did in the 1980s and 1990s. Seventh, set up a task force made up of representatives of the country's main labour unions and businesses, as well as government officials, to explore ways to ensure maximum participation of organized labour at the local, provincial, and national levels in the planning, building, running, and servicing of the new hydrogen energy infrastructure. Eighth, work with local and provincial governments, public and private utilities, and civil-society organizations to set up prototype distributed-generation associations and HEWs in the poorest urban and rural communities, with the goal of creating universal energy independence. Ninth, establish a working committee made up of provincial education ministries and the leading technical schools, colleges, and universities to design curriculum and provide course instruction to train the next generation of workers in hydrogen-related technical skills. Tenth, assemble an intellectual advisory board to address the many social, cultural, and political ramifications and consequences of making the transition from a fossil fuel–based economy to a hydrogen-based one.

The opportunity to make a fundamental change in energy regime, remake the underlying technological infrastructure, and spur wholly new types of commercial activity occurs only occasionally in history. This is one of those occasions. We find ourselves at the dawn of a new epoch in history. Hydrogen, the very stuff of the stars and our own sun, is now being seized by human ingenuity and harnessed for human ends. Charting the right course at the very beginning of the journey is essential if Canada is to make the great promise of a hydrogen age a reality for its children and a worthy legacy for the generations that will come after them.

A HISTORY OF ENERGY (CONTINUED)

1837   The astronomer John Frederick Herschel, while on an expedition to the Cape of Good Hope in South Africa, makes and uses a solar cooker or "hot box." The temperature inside the box reaches 116°C.

1839   William Robert Grove, a British judge and scientist, tries to harness the electricity created from the reaction of hydrogen and oxygen (reverse electrolysis) into a "gas battery." Dr. Francis Thomas Bacon (a direct descendant of the famous philosopher) will improve the device and create the first alkaline fuel cell in 1932. It will not be until 1959 that he develops a truly workable model.

1844   The American Charles Goodyear accidentally drops rubber mixed with sulphur on a hot stove and discovers what will come to be known as vulcanization, a process that strengthens rubber. His patented process means that rubber will no longer melt in hot weather and crack in cold, making it useful for a range of applications, including (eventually) automobile tires.

1851   The Irish-born scientist William Thomson (later Lord Kelvin) publishes his account of the second law of thermodynamics, resulting in a design for the heat pump, a device used to transfer heat between a low-temperature medium and a high-temperature one, the precursor for heating and air conditioning.

1854   Abraham Gesner of Halifax, Nova Scotia, formerly the Provincial Geologist of New Brunswick, the first government geologist in a British colony, immigrates to the United States and opens a plant in New York to manufacture kerosene using a patented process. The "Halladay Standard" windmill, invented by Daniel Halladay, is the first commercially successful windmill in North America. It pivots based on wind direction and regulates its own speed so as not to destroy itself by turning too fast.

1859   "Colonel" Edwin Drake strikes oil in Titusville, Pennsylvania, after drilling twenty-one metres down at the slow rate of one metre per day. He is soon steadily producing twenty-five barrels per day. Despite his role as the founder of the oil industry, Drake will die penniless.

▼   CONTINUED ON PAGE 126

05.

HYDRICITY, THE UNIVERSAL CURRENCY

GEOFFREY BALLARD

EDITORS' INTRODUCTION

If Jeremy Rifkin and Jules Verne are the evangelists of the hydrogen economy, Dr. Geoffrey Ballard and Sir William Grove are the engineers. After all, Verne's vision that water would one day be used as a fuel was based on the scientific work of Grove, and Rifkin's vision of the hydrogen economy depends on the pioneering work of Ballard. That's because at the heart of the hydrogen energy future is a simple device called the hydrogen fuel cell.

The story of the fuel cell begins a good thirty-five years before Jules Verne would write *The Mysterious Island*. The year was 1839 when William Grove, a young lawyer and scientist, invented the first fuel cell. After experimenting extensively on batteries, Grove decided to reverse the process of electrolysis. Sending an electrical current through water splits it into oxygen and hydrogen, so Grove theorized that combining oxygen and hydrogen would produce electricity and water. He built the first "gas voltaic battery" or fuel cell to test his thesis, and he was proved right.

When Grove died in 1896, so, in a sense, did the fuel cell. But thirty-seven years after Grove's death, the heir to the fuel cell was born. His name is, of course, Geoffrey Ballard. After growing up in Canada, Ballard moved to the United States to get his Ph.D. in geophysics from Washington University in St. Louis, Missouri, before joining the army. But the 1973 energy crisis was the turning point of his life. Realizing that the world needed to find a way to kick its dependency on fossil fuels, the young engineer quit his job with the government and started his own company. The engineer had a dream: to end the age of the internal combustion engine.

For seven years Ballard tried to pioneer a lithium-based superbattery, but he went bankrupt in the process. That's when he started getting interested in the fuel cell. The only people working on fuel cells at the time were at NASA, which had installed them in its Gemini orbiters to provide both fuel and drinking water for the astronauts. But even NASA could not find a way to generate enough power from the cells. Ballard decided he could do better.

Good technology, Ballard once told us, is accommodating, meaning that the more you demand of it, the more it complies. It doesn't get more and more complicated; rather, a good technology stays simple. With this in mind, Ballard kept experimenting with the fuel cell, trying to find a way to boost its power output. In 1979, he and his partners Paul Howard and Keith Prater founded Ballard Power Systems. The age of the fuel cell was born.

By the late 1990s Ballard Power Systems had convinced much of the world that the fuel cell was the future, or at least a serious option. Ford and DaimlerChrysler invested hundreds of millions of dollars. With the big money, however, came a new group of managers, and in 1997, after some infighting, Ballard retired as chairman of his company. But his dream was far from over.

For Ballard, the fuel cell was merely the beginning of the hydrogen revolution. The second phase has just begun. He calls this phase the Hydricity economy, and both the dreamer and the engineer in him are providing the ingenuity to make it happen. Let's see how the father of the modern fuel cell is trying to fuel the future.

MUCH OF THE FOCUSED THINKING concerning the degradation of our environment began more than a quarter of a century ago with books such as Rachel Carson's *Silent Spring*, but what really got everyone's attention — including my own — was the energy crisis of 1973–74, when the Organization of Petroleum Exporting Countries (OPEC) put an embargo on oil. Suddenly there were lineups at the gas pumps, oil prices rocketed upwards, and a general sense of panic prevailed. In America's national effort to find ways of solving the energy crisis, I was called to Washington, D.C., to work with a team tasked with developing an energy self-

FIRST, WE HAVE AN ENDURING LOVE AFFAIR WITH OUR CARS — OR, AS THE POLICY WONKS LIKE TO SAY, A DEEP SOCIO-ECONOMIC DEPENDENCE ON PERSONAL TRANSPORTATION — AND WOE BETIDE THE PERSON WHO INTERFERES WITH OUR ADORATION.

sufficiency budget for the United States. It was fascinating work, and I came away from that experience six months later with a radical but unshakable conclusion: if planet Earth was to survive as a healthy, living organism, our dependence on fossil fuels and the internal combustion engine would have to end.

As a response to that conclusion, I built a company based on technologies that could replace the internal combustion engine: first the battery, and then the hydrogen fuel cell. But it took me another eight years of research to learn two fundamental truths about the automobile. First, we have an enduring love affair with our cars — or, as the policy wonks like to say, a deep socio-economic dependence on personal transportation — and woe betide the person who interferes with our adoration. And the second truth explains the first: the hold that the internal combustion engine has on society comes from the fact that this piece of technology allows the scaling of power and energy separately. Let me explain what this means, because understanding it is crucial. Engines and gas tanks are separate from each other, so they let us choose just how our cars perform. If we want more power, we put in a bigger engine; if we want to go farther on a tank of gas (more energy), we put in a bigger gas tank; or we choose any combination of engine and gas-tank size we need in order to meet the "mission profile" of the targeted sale.

Batteries and other suggested devices to replace the internal combustion engine do not have this versatility of allowing the power and the energy to be scaled separately. Instead, the ratio of power to energy is fixed by the choice of device, which then dictates limits to the speed, size, or distance travelled. I realized that to replace the internal combustion engine we needed something that could scale power and energy separately, like the internal combustion engine can, but without releasing so many pollutants. The only technology that I believe is available today to do this is the proton exchange membrane (PEM) hydrogen fuel cell.

The PEM fuel cell — the fuel cell usually considered for the automobile — is actually a rather simple piece of technology. The cell consists of a plastic membrane, not unlike a thick piece of plastic wrap, coated on either side with a platinum catalyst that promotes the desired chemical reactions. When air is brought to one side of this plastic membrane and hydrogen to the other, the electrons in the hydrogen atoms are pulled off and the hydrogen nuclei that are left behind migrate through the plastic. On the other side of the membrane, the hydrogen and the oxygen in the air join to make pure water, while the stripped-off electrons go around the outside of the membrane, forming an electrical current. It is this small electrical current that we use to power the motors in our fuel-cell cars. To produce useful voltages, or more power, we stack these cells together — just like you would place batteries end to end in a flashlight — to form what we call a fuel-cell stack.

It took years to perfect a way to output enough power from a fuel cell using inexpensive materials, but when we finally did we knew the fuel cell was the way of the future. My partners and I

HYDROGEN IS NOT, TECHNICALLY, AN ENERGY SOURCE BUT A CURRENCY THAT MAKES ALL SOURCES OF ENERGY AVAILABLE TO THE ENTIRE ECONOMY.

built Ballard Power based on the fundamental idea that since the fuel-cell engine has the same freedoms of scaling as the internal combustion engine but without the detrimental effects of pollution, fuel-cell cars are poised to be our next enduring love.

The fuel cell not only solves the scalability problem but also actually allows us to transform our economy from one based on fossil fuels to one based on any primary source of energy. Whereas a petrochemical economy is single-source dependent on crude oil, a hydrogen economy can utilize all available primary sources of energy. This transformation will have dramatic implications for the way our society functions, and the key moment will be when hydrogen and electricity become so indistinguishable from each other that we will be able to refer to them as a joint currency, which I call hydricity. Let's explore what that means.

Hydrogen is not, technically, an energy source but a currency that makes all sources of energy available to the entire economy. Currencies are exactly what their name implies. Just as you get paid in a currency such as dollars, pounds, or yen for your work, so nature's sources of energy are transformed into currencies such as gasoline, hydrogen, or electricity. These currencies are then moved to the locations where energy is needed and where they are expended — or, let's say, where they are spent — in the same manner as you take your pay and buy food or shelter or transportation. The energy currency lets us bring the energy from the source to the place of need.

The most common energy currency is electricity, because most natural sources of energy are used to generate electricity. We burn coal, oil, gas, and wood to generate electricity, and we dam rivers and split atoms to generate electricity. Then that electricity is transported to a place of need — your home, say, or a factory — and converted into any form of service that's required — your water is heated, your lights shine, your computer runs. In fact, it's fair to say that the developed world is not so much hooked on oil as it is hooked on electricity. Look at the data: as a major conservation attitude kicked in during the seventeen years after the oil (energy) crisis of 1973–74, there was actually a per-capita decrease in the total consumption of energy in North America. But not all forms of energy. During the same seventeen years there was a 28-percent increase in the per-capita consumption of electricity. In other words, North America, and the rest of the world, realized that it needed to find a better energy source than fossil fuels to produce electricity.

But electricity has a drawback: when it is generated, it must be either used or stored. Until the fuel cell, there has been no easy, economical way to store electricity. Once the electricity is generated, it has to be used immediately or it goes to waste. That's why our electrical generation system is designed for peak needs rather than for average needs, which has resulted in excessive over-capacity at slack usage times.

Let's put some numbers on these seemingly vague terms of "peak," "average," and "slack," to help us visualize the full impact of what it means to be able to store electricity. I'll choose a typical late-spring day in California as an example. The usage patterns are very similar the world over, though they do become exaggerated in the wintery North as home heating is required and again in the summery South when air-conditioning is used. But, in my opinion, the example is fair and does not hide any important truths.

On this average spring day in California, from midnight to 6 in the morning only about 35 percent of the state's generating capacity is needed; the other 65 percent of the generating capacity is turned off and unused. From 6 to 9 in the morning the demand rises to 88 percent of capacity and holds fairly steady until a dip in mid-afternoon, when it drops back to 75 percent of capacity. Between 4 and 6 in the afternoon the demand rises quickly to 100 percent of the generating capacity, before dropping steadily back to 35 percent by midnight. If an exceptional heat wave or cold snap persists, California will experience "brownouts" as the demand exceeds the state's ability to produce power.

This daily "peak" and "slack" dynamic could be changed if we could store electricity. So how could we store it? Well, before the advent of the fuel cell, the best way to store electricity was in a battery. The battery is not commercially viable for storing large amounts of energy, however, and does not allow us to scale power and energy separately. The hydrogen fuel cell has changed the way we can set up and use our power-generating plants because it solves the electricity storage problem. Instead of having to use the electricity immediately, we can store it as hydrogen, which can be used later. In other words, with the fuel cell we can now afford to produce hydrogen using electricity during off-peak times, store the hydrogen, and then feed it back through a fuel cell to generate electricity at peak-need times. This is often referred to as "peak-shaving." The fuel cell allows us to peak-shave and plan ahead for electrical usage. Using this method we would be able to meet the electrical demand with about 75 percent of the generating capacity we now require.

But is this economical? Surely there is a lot of waste. Yes, there is waste, but peak-shaving is still economical. The round trip — producing hydrogen at slack times by electrolysis and generating electricity from hydrogen with a fuel cell at peak times — is a three-to-one proposition. In other words, you only get back one-third of the electricity you used to make the hydrogen. From an economic point of view this is worthwhile, however, because the electricity at off-peak times costs about 3 cents per kilowatt-hour while at peak times it can cost 35 cents per kilowatt-hour. That is a ten-to-one proposition and very profitable, apart from the fact that it works against brownouts and reduces the amount of generating capacity needed. Those are also not bad numbers when you consider that for the average automobile only 18 percent of the energy in gasoline gets through to the rubber on the road. Incidentally, if California used its off-peak generating capacity to produce hydrogen, the off-peak capacity — currently unused — would be enough to fuel one-third of the automobiles in the state.

All this is an example of what I call the hydricity economy, that is to say, the result when the hydrogen economy, which uses hydrogen as a currency instead of gasoline, combines with and utilizes electricity, civilization's present currency of choice. A fuel-cell engine in a car, for example, utilizes hydrogen as its currency to generate electricity, but it also uses electricity as a currency because the electricity generated by the fuel cell is then sent to an electric motor in the car to make it move. A fuel-cell vehicle is really an electric vehicle.

To see the potential of the hydricity economy, let's look first at the energy system currently in place for the transportation

YOU CAN ACTUALLY BEGIN TO CONCEPTUALIZE THE AUTOMOBILE AS A POWER PLANT ON WHEELS.

sector. The first thing we notice is that the system is single-source dependent; it depends on reforming crude oil into gasoline for the internal combustion engine. I don't need to go into the international political situations, wars, and threats of war that have emerged from our inability to find a suitable substitute for crude oil. None of us need to be told this story again. Suffice to say, the oil economy, for all the many benefits it brings us, is expensive and dangerous to maintain, is geopolitically volatile, and has all sorts of negative consequences on our health and well-being.

The fuel cell changes this dynamic. First, as I mentioned earlier, it's not single-source dependent. Any primary source of energy can be used to make electricity, which can then be stored using hydrogen. And second, the car is transformed from just being an energy consumer to also being an energy producer. You can actually begin to conceptualize the automobile as a power plant on wheels. This means that the automobile is a natural link for distributing power. That might not be the case for the first generation of fuel-cell vehicles, but certainly the second or third generation will be what we call "regenerative fuel-cell vehicles."

These will have "vehicle-to-grid" (V2G) capability. Let me explain what these terms mean.

Since today's technology allows us to receive and store hydrogen on board a fuel-cell car, it also allows us to generate our own hydrogen on board the car when it's plugged into the electricity grid either at home, at the office, or in a parking lot. A car with this ability to refuel itself is called a regenerative fuel-cell vehicle. If we also provide data and electronic connections when we plug the car in to refuel, it can be turned on by the building or lot to which it's connected. In this way the car can provide electricity to the grid. This is V2G capability.

The implications of this configuration are enormous, when you consider that 85 percent of the vehicles we use are parked either at home, at work, or in a parking lot, even during rush hour. For example, there are approximately 25 million automobiles on the road today in California. Most automotive manufacturers plan on installing 50- to 100-kilowatt engines in their average automobile. If each automobile can generate 50 kilowatts, then a single vehicle can generate enough power for five to ten homes. One hundred fuel-cell vehicles can generate over 5 megawatts, more than enough to power a fifty-story office tower. Twenty thousand fuel-cell vehicles connected to the grid represent the power saved in California by the rolling blackouts. One million vehicles, or 4 percent of the cars registered in California, represent that state's total stationary generating capacity.

There are countless possibilities for using this amount of distributive generating capacity on and off the road. Let's say you are going fishing. Everything you bring could plug into your car: your computer, portable stove, heater, TV, reading lights, you name it. Let's say you park at the office. You could negotiate a free parking space by trading the office management the ability to turn on your car and use 25 percent of the on-board hydrogen; then the office has uninterrupted power even if the grid fails. The office can dispense with a separate uninterrupted power system and charge more for its work space. Let's say you have a cottage and the grid is twenty miles away. No problem, just plug in your car when you get there and the cottage has power.

It is impossible to avoid the conclusions that most of the power-generating capacity of a developed nation is in its rolling

IT IS IMPOSSIBLE TO AVOID THE CONCLUSIONS THAT MOST OF THE POWER-GENERATING CAPACITY OF A DEVELOPED NATION IS IN ITS ROLLING MOTOR VEHICLE STOCK, AND THE ADVENT OF THE FUEL CELL MAKES THIS HUGE POWER-GENERATING CAPACITY AVAILABLE FOR UNRESTRICTED DAILY USE.

motor vehicle stock, and the advent of the fuel cell makes this huge power-generating capacity available for unrestricted daily use.

But the impact of the hydricity economy goes beyond transportation. Domestic security and energy security take on an entirely different flavour for a country with a hydricity economy. Today we face the insecurity of a few terrorists with hiking boots and backpacks of plastic explosives being able to drop key power-grid links in the vast wilderness our grid traverses. With a hydricity economy power could be restored almost instantly by plugging in a small percentage of our cars. Today we face dependence on oil, the global supply of which is at the whim of nations with very different, often violently opposed, values to ours. With a hydricity economy we could free ourselves from fossil fuel dependence and the influence of volatile regions such as the Middle East.

What will drive us to move more rapidly towards the hydricity economy? Will it be the fact that, as some people suggest, we are running out of oil? A number of recent studies in the

United States and Europe start with preambles that express concerns with the supply of petroleum. I hold no such fears. Dr. Hans-Holger Rogner, head of Planning and Economic Studies at the International Atomic Energy Agency in Vienna, shows clearly that there is at least 200 years of petroleum available even under very pessimistic circumstances. We should not be changing the energy system because of a fear of limited petroleum reserves. We should change the energy system because the current system undermines energy security and unacceptably destroys the Earth's atmosphere. And the primary reason, in my opinion, to reduce our reliance on carbon-based fuels is that we are sickening our children with inner-city pollution.

Air pollution takes many adverse forms, but the worst one, to my mind, is the foul atmosphere that we inflict on inhabitants of the inner city. This has been widely ignored in the environmental debates, where the arguments are directed to cleaning up the upper atmosphere, ozone holes and depletion, and warming trends that could inundate coastal areas. I acknowledge that the atmospheric conditions can change drastically with only minor changes in one parameter and we could trigger the events that would produce catastrophic changes such as shutting down the Gulf Stream. However, the down-home truth is that burning fossil fuels is killing people today and this is not some probabilistic fear about the future. Millions of people throughout the world are being sickened and killed by bad air. The World Health Organization attributes 700,000 annual deaths to air pollution.

The evidence relating to the immediate health impacts of urban pollution are mounting. The journal *Nature* reports that if the coal-fired power plants of the Midwest United States reduced their emissions, there would be a reduction of 3 million sick days and 10,000 asthma attacks annually. A ten-year study funded by the California Environmental Protection Agency's Air Resource Board concludes that inner-city smog can cause asthma in children. We have known for some time that smog can trigger asthma attacks, but this study is evidence that smog causes the asthma itself. Scientists at the University of British Columbia and St. Paul's Hospital in Vancouver say that air pollution caused by automobile emissions can now be added to the known risk factors for strokes and heart attacks.

Maybe the answer is not to make a radical switch to hydrogen, but to follow the regulations of the Kyoto Protocol? In my opinion, Kyoto is not the answer at all. In fact, it's part of the problem. The Kyoto Protocol is a conservation strategy. In other words, it's a "use-less" strategy, a band-aid approach to a massive hemorrhage. I believe that no developed nation that has seriously

FOR SOCIETY TO CONTINUE ITS PROGRESS IN MEDICINE, SOCIAL RESPONSIBILITY, SCIENCE, EDUCATION, AND GENERAL QUALITY OF LIFE, WE MUST ENSURE THAT THERE IS AN EVER-INCREASING SUPPLY OF ENERGY PER CAPITA.

studied the environmental issues confronting us can in good conscience sign this protocol. I believe that nations attempting to implement the Kyoto Protocol will take a huge step backwards. I believe that the protocol says "business as usual," and that it will encourage governments to do some very creative bookkeeping to mask unwillingness to face the real issues.

To quote from a *National Post* editorial on December 19, 2002, "Kyoto requires that Canada reduce its emissions by 30% vis-à-vis 2002 levels. The probability that Ottawa will reach this goal without using bookkeeping tricks is zero."[1]

Economic progress, as we know it, correlates very well with per-capita energy consumption. So do all other forms of social progress. For society to continue its progress in medicine, social responsibility, science, education, and general quality of life, we must ensure that there is an ever-increasing supply of energy per capita. With human populations still on the rise, progress will not be sustained if we attempt to stabilize, let alone reduce, our energy production in order to reduce the emissions of the current energy-source mix. We must increase our supply of energy, not reduce it. The current energy mix, however, is too heavily weighted to the fossil fuel sources to allow us to increase energy production the way we have in the past. Only by moving to the hydricity economy can we meet rising energy needs without suffering from the negative consequences of fossil fuel emissions.

Broadening our thinking from our North American perspective, note that only 12 percent of the world population have access to some form of motor-power transportation. There are currently six billion people on the Earth, with a population of ten billion expected later this century. Most of these people have, or shortly will have, access to global communications through television satellite links. All of these people are becoming increasingly aware of the high standard of living of the citizens of developed countries. In more prosaic terms, every one of the "have-nots" will soon see just what the "haves" have. They will even get a distorted picture in the worst of all possible directions: they will see North American sitcoms as if they represented the way the average North American lives. To quote from public testimony from General Motors, "Most of these people are young, globally aware, web-connected, and residing in economies with escalating demands for personal transportation."[2] One doesn't have to resort to statistics to recognize that if 12 percent of us can cause such havoc to the Earth's atmosphere, the other 88 percent will spell out a doomsday for the planet if they follow our examples of energy source and use.

What does all this mean? In the long term, as I've said, we need to limit, then reduce, the use of coal and petroleum, the carbon-based fuels, in satisfying our energy needs. But that will take time. In the short term, it may be necessary to maintain, and perhaps even increase, our coal- and petroleum-based energy production, but this should be done with great care, utilizing the latest technologies. For example, a great deal of applied research has been done in recent years to decrease the emissions from coal-fired power plants. The resulting techniques are collectively grouped as

IT'S QUITE CONCEIVABLE THAT YOU COULD BUY A CAR, SIGN A UTILITY AGREEMENT FOR A FEW HOURS' USE A WEEK, AND GET FREE FUEL FOR THE LIFE OF THE CAR.

"clean coal technologies." Apart from these techniques, greenhouse gas emissions can be sequestered back into the earth and the oceans and prevented from entering the atmosphere. Again, all these are band-aid solutions until we move more dramatically to the hydricity economy.

Of course, I've painted a picture with very broad brush strokes, and such broad treatment leaves many questions in the minds of thinking people. I can't anticipate or answer all these questions, but some of the issues that recur deserve a brief explanation.

First, for the hydricity economy to really take shape, there would have to be an unprecedented change in the financing of energy utilities. The majority of the generating capacity represented as rolling stock would have been paid for by the fuel-cell vehicle purchasers. Utilities would negotiate with car owners for use of this generating capacity when it was not needed for transportation. There would have to be limitations on use and methods of credit and payment, but in general this capability has already been developed by the credit-card industry. All a car would need is a data port and an agreement for use. It's quite conceivable that you could buy a car, sign a utility agreement for a few hours' use a week, and get free fuel for the life of the car. If you can get free fuel for the life of your car, this will change the dynamic of where you live with respect to your job because you're no longer stuck with balancing land cost against commuting costs for employment.

Second, a question often arises about how we will switch from gasoline stations to hydrogen stations. The corner gasoline station will eventually disappear. With 850 million vehicles on the road today and 50 million new vehicles a year being manufactured, it will be years, however, before the increase in gasoline usage begins to diminish. Therefore I doubt that the gasoline stations of today will be the hydrogen stations of tomorrow. It seems much more likely to me that the parking lots of major retail outlets such as Wal-Mart will be the hydrogen outlets that supplement the home refueling capability that comes with regenerative fuel-cell vehicles.

Another question that often comes up is, Isn't hydrogen dangerous? What about the *Hindenburg*? I would answer in several ways. First, when we put fuel-cell buses filled with hydrogen on the streets of major cities such as Chicago, people lined up to get a ride. No one really worried whether hydrogen was dangerous. So the general perception is not one of concern. Second, hydrogen has been studied for many years because it's a major commercial product used in a number of industries from oil refining to margarine production. Tank cars of hydrogen sit in freight yards across North America and travel daily by rail and tanker truck through most major cities. There are few incidents of concern and fewer accidents. Third, from a technical point of view hydrogen is usually considered safer than gasoline. Gasoline is heavy and pools when ignited, whereas hydrogen is light and rises into the air when lit. Fourth, the *Hindenburg* disaster was about flammable paint and burning diesel from the engines much more than it was about hydrogen. But the perception of danger remains, and much education and experience will be needed to completely allay the fears that people have. I have no doubt that hydrogen will become very acceptable as people weigh both its advantages and their own concerns.

What are the technical roadblocks to implementing the hydrogen economy? Why aren't fuel-cell cars on sale today? There are some major technical hurdles to overcome before hydrogen can move front and centre as our fuel of choice and fuel-cell vehicles can grace our roadways. One problem is that hydrogen is light and bulky. We have not yet found the ideal way to store hydrogen on board a vehicle. High-pressure gas tanks are bulky and costly. Metal hydride storage is heavy. Liquid hydrogen, preferred by many engineers, has problems of long-term storage and safety.

Another hurdle is cost. The fuel-cell engine is an order of magnitude more expensive than the internal combustion engine and drivetrain. The fuel-cell engine is at least in the $500-per-kilowatt price range, while the internal combustion engine is in the $50-per-kilowatt price range. Economies of scale can reduce the cost of fuel-cell engines when they are mass-produced, but mass production will at best only cut the cost in half. Fuel-cell engines have a long way to go to become price competitive, but price competitiveness will come because the basic ingredients of the fuel-cell engine lend themselves to inexpensive production.

As well, fuel cells still have some difficulty with reliability. The fuel cell needs to be assured of 5,000 hours of operation before failure, and few fuel-cell manufacturers are willing today to warrant this performance.

There is also the classic "chicken and egg" problem with the introduction of the fuel-cell vehicle. It is impossible to sell large numbers of fuel-cell vehicles if there is nowhere to fill them up with hydrogen. On the other hand, oil companies are reluctant to spend millions of dollars to add hydrogen refueling to their gasoline stations if no fuel-cell vehicles are on the road needing to be refueled. Given this chicken and egg situation, how do we jump-start the hydrogen economy?

Massive changes in human behaviour are not brought about with legislation. It doesn't seem to work. We have legislated high-occupancy vehicle lanes on the highway and they are empty. I really believe in an economic basis for change, without massive government support or tax rebates or any other legislation. At my new company, General Hydrogen, we are saying, "Okay, you want to introduce the hydrogen economy? You want distribution channels and things like that? Well, who can benefit today, with the present cost structure of fuel cells? Who has a need that is sufficient? Who has a market niche where it pays to enter the hydrogen economy at today's fuel-cell prices?" And market niches do appear to be available.

For example, I see the early-stage applications as being anything where a lead-acid battery can be replaced. Battery-powered forklifts, for example, can run much more efficiently on hydrogen. They are very heavy; right now they run out of power in four hours. But you can run them on fuel cells for twenty-four hours and refuel them in five minutes. I think there are sufficient productivity gains to compensate for the initial changeover costs. From material-handling equipment the fuel-cell engine can migrate into delivery trucks and then to the retail outlets, and from there to the parking lots to bring in customers. As sales and production increase, the price will slowly come down the cost curve. At each price break, new markets will open up, until the price and volumes are suitable for the automobile.

I have left one of the most important questions until last. This question leads to a very controversial answer, yet it's an answer I believe we must face head-on. The question goes something like this: "You've said that hydrogen is just a currency, that it's not a primary energy source; you've said that the Kyoto Protocol is a conservation strategy you reject, that the world must increase its use of energy. Where will all this energy come from to make all the hydrogen your hydricity economy needs?" The answer, in a word, is *nuclear*.

Hydrogen is only a currency, but it's a currency that allows any primary energy source to be utilized. In Iceland the primary energy source will probably be geothermal, in Sweden it may be hydroelectric, in Canada it may be a combination of hydroelectric and nuclear, and in Argentina it may be wind. Throughout the world, many remote locations will employ solar sources. But environmentally desirable as some of these renewable sources of energy are, they are unlikely to provide the vast amounts of primary energy that social progress will continue to demand.

If carbon-based energy sources must be set aside, and I believe they must, then the only remaining viable source, at this stage of our technical development, is nuclear. Yes, there are other possibilities in the future. Recently there has been speculation in the press that hydrogen could be mined directly from deep Earth sources, for example. But within the scope of today's technology, nuclear fission is the only viable, clean source of large quantities of energy. Nuclear-generated electric power does not pollute our atmosphere, and mitigates global warming.

With this focus in mind, a nation's entire energy system can gradually shift away from dependence on oil and petroleum without a massive disruption of the economy. It's quite possible for the developing world's populations to adopt the fuel cell and the hydrogen economy and skip the fossil fuel infrastructure that we have in the West, and then utilize the ability offered by these technologies for a distributive power network instead of a hard-wired power grid. In other words, we can control the evolution of nations' energy utilization into directions and systems that are appropriate for each nation and are least disruptive to our global ecology.

With the hydrogen economy we have choices. Any primary energy source can be used to produce electricity. Electricity can produce hydrogen. And hydrogen can be stored, shipped, and used to again make electricity. This interchangeability will assure that electricity and hydrogen are our currencies of choice. Hence hydricity.

A HISTORY OF ENERGY (CONTINUED)

1862 A German travelling salesman named Nicolaus August Otto builds his first four-stroke internal combustion engine, which runs on gas rather than steam. The appropriately named "Otto" engine will not be patented until 1877, when he finally perfects the ignition mechanism. The Otto engine will make the development of automobiles, motorcycles, and eventually airplanes possible.

1864 The Scottish physicist James Clerk Maxwell begins work on his *Dynamical Theory of the Electromagnetic Field*. His theories, which will be posthumously distilled into the Maxwell's Equations, will be considered by some to be precursors to the twentieth-century theory of relativity and the quantum theory. Maxwell's theories will also lead to the development of electric power, radios, and televisions.

1867 After years of experimentation with explosives, the Swedish chemist Alfred Nobel discovers a way to absorb nitroglycerine into *kieselguhr*. The result, which he names dynamite after the Greek word for power, proves safer to handle than earlier explosives. His patents will help to make him one of the wealthiest men in Europe. After he dies, it will be discovered that he has left much of his estate to a fund, the interest of which is to be given to those who have benefited mankind, as the Nobel Prizes.

1868 Standard Oil is founded and organized by John D. Rockefeller. By 1878, Standard Oil will control 95 percent of the U.S. oil business.

1870s The chemical engineer Herman Frasch, a contractor for Standard Oil, develops a process to extract sulphur from oil using various metal oxides and thus to reduce the unpleasant smell of sulphur. This allows Standard to develop its Lima-Indiana oil field, the oil from which has an exceptionally high sulphur content.

1874 Jules Verne imagines hydrogen being used as a fuel in his novel *The Mysterious Island*.

1875 John Macoun, a botanist with a Geological Survey of Canada expedition, reports "tar mingled with sand" near the confluence of the Clearwater and Athabasca Rivers.

▼ CONTINUED ON PAGE 152

06.

IS NUCLEAR ENERGY THE ANSWER?

ALLISON MACFARLANE

EDITORS' INTRODUCTION

Late in the day on April 25, 1986, the operators of the Chernobyl nuclear plant began to shut down power to their reactor 4. The operation was intended to be a routine study of the plant's ability to continue to run following a loss of electricity. The Chernobyl power station consisted of four power reactors and a generating turbine, situated by Pripyat, just north of Kiev. As the plant turbine generator shut down, power was reduced to 1,000 megawatts, where it was supposed to stay. However, the power continued to drop due to operator error, until it reached 30 megawatts. This started an unexpected and calamitous chain reaction.

As the electrical supply was reduced, the flow of cooling water began to slow, causing the temperature inside the reactor to rocket upwards. This in turn caused a buildup of steam in the cooling channels. Moments later — at 1:23 a.m. on April 26 — the reactor exploded, blowing off the steel roof and spraying fifty tonnes of nuclear fuel and graphite into the air. As the burning debris fell, it ignited a further thirty fires around the plant.

In an effort to put out the fire and limit the release of radiation, around 5,000 tonnes of fire-retardant material was dropped onto the burning reactor by helicopter. And then, to stave off an imminent nuclear meltdown, the plant was cooled by adding twenty-five tonnes of liquid nitrogen a day to the earth around the plant to prevent the temperature inside the core from rising any further.

In his book *The Truth About Chernobyl: Heroes and Villains of the Chernobyl Disaster*, Russian nuclear engineer Grigori Medvedev estimates that the accident at Chernobyl released the equivalent amount of radiation as ten atomic bombs of the type that was dropped on Hiroshima. Some of the human costs of the accident are known, like the twenty-eight firefighters who died from direct exposure to the fires and radiation. But the indirect costs are harder to quantify. Thousands of children — perhaps as many as 10,000 — contracted thyroid cancer as a result of exposure to high levels of radiation. And over 350,000 people were permanently evacuated from surrounding areas.

For many of us, the possibility of accidents like Chernobyl still underline our inherent fears concerning nuclear power. It does not help that nuclear energy is tied so directly to the still very real threat of nuclear weapons, a fear that has been amplified by the terrorist attacks of September 11, 2001.

But for all this, Geoffrey Ballard, the man who pioneered the hydrogen fuel cell, believes that nuclear energy is the best and the only really viable wide-scale source of the hydrogen required to power a hydrogen economy. As he told us, the risks of a recurring accident like the one at Chernobyl — which by all accounts are extremely slim — have to be weighed against the risks associated with increased uses of fossil fuels.

With Ballard's vision of the future on the line, perhaps it's time to take another look at nuclear power. And few are better equipped to do that than Allison Macfarlane. While studying and teaching at institutions such as MIT's Security Studies Program, George Mason University, Radcliffe College, Harvard University, and Stanford University, she has focused on international security and environmental policy associated with nuclear weapons and nuclear energy. She is currently Associate Professor of International Affairs and Earth & Atmospheric Science at Georgia Tech in Atlanta, Georgia.

IMAGINE YOUR NEIGHBOURHOOD with a nuclear power plant a short drive away. You can see the tips of the cooling towers from your upstairs window. You know the plant is a source of clean, greenhouse-gas-free (or almost-free) energy production. It is a new, "inherently safe" designed plant, built with subsidies from the federal government, your taxpayer dollars. The waste from the plant is kept on site in a reinforced bunker — that and the armed-guard force protect it from sabotage. The waste will eventually go to a repository located within 800 kilometres of your home. In fact, most communities have similar nuclear power reactors; yours is not unique. They are helping to defer the deleterious effects of greenhouse gases on the climate. But are you comfortable with it there?

Worldwide electricity demand is expected to increase in the years to come, not only from familiar residential and industrial uses but also potentially from new demands such as the formation of hydrogen fuel. Given a business-as-usual scenario of the future, where the world's population increases to about ten billion by 2100, electricity generation is expected to almost double by 2020, quadruple by 2060, and quintuple by 2100. In concert with the growing evidence of a link between fossil fuels and climate change, a decrease in reliance on fossil fuels for electricity supply is needed. Estimates suggest that by 2020, carbon emissions from electricity

generation alone will contribute double that of the current rate. If
these levels are not reduced, the associated climate warming may
produce disastrous results. How can we meet the electricity needs

THE QUESTION IS, CAN NUCLEAR STEP INTO THE VOID AND ACTUALLY EXPAND ENOUGH TO TAKE A BIGGER BITE OUT OF THE CARBON LOAD?

of an evolving world but reduce carbon emissions? Nuclear power
may provide the answer.

THE PAST, PRESENT, AND FUTURE OF NUCLEAR POWER

Currently, 16 percent of the world's electricity is supplied by
nuclear power. Certainly the current nuclear capacity reduces the
atmospheric load of greenhouse gases that would otherwise be
generated by coal-, oil-, or gas-burning plants. The question is,
can nuclear step into the void and actually expand enough to take
a bigger bite out of the carbon load?

First, it is important to understand where nuclear power is in
relation to other sources of electricity. As of March 2003, there
were 437 commercial nuclear power reactors in thirty-one countries, with a total of about 360 gigawatts capacity, enough to power
400 million California homes. In comparison, fossil fuel plants had
a worldwide capacity of about 2,400 gigawatts at this time, enough
for almost 3 billion California homes. Though the United States
gets only 19 percent of its electricity from nuclear power, it still has
the largest nuclear energy fleet in the world. As of 2003, the United
States had 103 licensed operating reactors in thirty-one states.

Much has been made of the potential for expanding nuclear power, but this would only occur over the long term. Nuclear capacity has hardly grown in the last fifteen years and will not likely grow much over the next fifteen years. The United States, for instance, last ordered a new nuclear power plant in 1978, and no plant ordered after 1973 was built (in large part, this

NUCLEAR CAPACITY HAS HARDLY GROWN IN THE LAST FIFTEEN YEARS AND WILL NOT LIKELY GROW MUCH OVER THE NEXT FIFTEEN YEARS.

was due to the accident at Three Mile Island). Thirty-three new nuclear plants are being constructed in eleven countries, predominantly India, China, Ukraine, Russia, and Japan. Little expansion is occurring in Europe and none in the United States or Canada.

In the United States and some European countries such as France, the trend is to extend the lifetime of existing reactors. The U.S. Nuclear Regulatory Commission (NRC) has begun granting twenty-year licence extensions for power plants. Germany, Sweden, and Belgium, on the other hand, have voted to phase out nuclear power before all plants have completed their licensed lifetimes. Furthermore, Austria, Denmark, Greece, Ireland, Italy, and Norway have prohibited the use of nuclear power. At the same time, many nuclear power plants in the United States and Europe have had power uprates, which have allowed the plants to increase their electricity generating capacity. These power uprates enable the plants to produce electricity more competitively.

Deregulation of energy markets in the United States and

elsewhere has allowed the restructuring of electric utility companies. For nuclear power, that has meant consolidation. In the U.S., for example, in 1991, 101 utility companies owned 110 reactors, but by 1999, the number of owners had decreased to 87, of which 12 owned more than 50 percent of the total generating capacity. Restructuring of nuclear utilities allowed larger companies to take advantage of economies of scale in maintaining and operating their plants. Previously, a company that owned a single nuclear power station may have decided to shut down instead of investing to replace aged, expensive equipment, whereas larger power plant owners can absorb these costs more easily.

The short-term future of nuclear power in countries like the United States looks somewhat brighter than it has in years. The regulatory climate is more favourable than before, and existing plants are more economical and therefore more competitive with fossil fuels, because the capital costs of building the nuclear power plants have been paid or the costs shifted to ratepayers. Finally, there are new reactor designs on the horizon.

But the future is not clear. Currently, all major expansion is occurring in developing countries, and even that is limited. Nuclear power is stagnating or shrinking in most developed countries. But nuclear power is the only currently existing, large-scale, geographically unlimited source of greenhouse-gas-free (or at least reduced) electricity. And for those countries that do not have cheap access to fossil fuel resources, nuclear power may be one of the only sure sources of reliable electricity. What would it take to expand nuclear power on a large scale, and what scale of expansion would reduce climate-change effects?

CAN NUCLEAR POWER REDUCE GREENHOUSE GAS EMISSIONS?

Nuclear power will not be the only answer to our energy needs in the future. First, nuclear power can only reduce greenhouse gas emissions from electricity production — not from fossil fuel use in transportation, for example. Second, the claim that nuclear power produces no greenhouse gas emissions is not actually correct. Greenhouse gases are emitted during the extraction of uranium for fuel, as well as during uranium processing and enrichment. Greenhouse gases are also emitted in the production of construction materials for nuclear power plants, such as concrete and steel. In addition, there are some minor emissions of greenhouse gases during reactor operations by secondary generators that are required in case of accidents. These generators must be tested on a regular basis, and during the testing they emit carbon dioxide and other gases.

Most nuclear power reactors in operation in the world today — those termed light water reactors — require uranium to be enriched in one of its naturally occurring isotopes, uranium-235, for use in fuel. The enrichment process can be very energy intensive, depending on the exact method used. The most energy-intensive enrichment process is gaseous diffusion, where uranium hexafluoride gas is passed through membranes that retard the movement of the heavier uranium-238 isotope. These plants tend to be huge as well as energy intensive. The Paducah, Kentucky, gaseous diffusion plant gets most of its electricity from coal-burning power plants. Thus, its annual operation creates the emission of about as much greenhouse gases as that from three 1,100-megawatt coal plants.

Other enrichment technologies, such as centrifuge plants, are less energy intensive than gaseous diffusion; nonetheless, they still use electricity. Unless a system can be made in which all the electricity used in uranium mining, milling, and enrichment comes form nuclear power itself, nuclear-produced electricity will result in the emission of some greenhouse gases.

The larger question that we are trying to address is, what is a reasonable expectation for greenhouse-gas-emission reductions from nuclear power? Today's nuclear power plants save 600 million tonnes of carbon per year from going into the atmosphere from equivalent power-producing coal-fired plants. But this is just 10 percent of the total amount of carbon released into the atmosphere each year. Nuclear power saves even less, about 250 million tonnes of carbon per year, for gas-fired power plants. For nuclear power to make a substantial dent in carbon emissions, it would have to reduce emissions by at least a third. By 2100, given that electricity usage may have increased fivefold, nuclear power would have to increase its share of production from one-sixth (what it is today) to one-third multiplied by five, or about ten times.

WHAT DOES EXPANDING NUCLEAR POWER MEAN?

What does a tenfold increase in nuclear power generation translate into?

This would entail building in the order of 3,200 new mid-sized nuclear plants worldwide, 900 of them in the United States alone. The implications of such an expansion are impressive in terms of the waste and nuclear weapons materials produced. If these 3,200 new plants were of the same variety as those that now exist in the United States, namely light water reactors, then the amount of used nuclear fuel produced on an annual basis would be 72,000 tonnes. That amount is equivalent to that planned to fill the entire U.S. repository at Yucca Mountain, Nevada. One percent of that fuel, 720,000 kg, would be plutonium, of which only 4 kg is needed to make a nuclear weapon.

What does this mean overall? First, it is important to note that I am only considering long-term scenarios here, where the tenfold increase takes place between 2050 and 2100. It would not be possible to increase capacity so greatly over the short term, within the next ten or twenty years. It would simply take too long to build all the nuclear plants and supply all the equipment and personnel.

Over the long run, however, it would be possible to complete such an expansion. In doing so, a number of significant issues would require resolution before the world turned to nuclear power as a (partial) solution to greenhouse gas emissions. These include the cost of building new power plants, safety issues posed by the existence of over 3,000 plants, huge volumes of nuclear wastes, nuclear weapons proliferation, and the potential for terrorist strikes on nuclear power plants. Other issues include public acceptance of nuclear power and the low-level doses of radiation it imparts, and infrastructure issues, especially those of an aging workforce with few replacements and a decreased manufacturing capability.

OBSTACLES TO NUCLEAR POWER EXPANSION: COST

Whether nuclear power is an attractive alternative to governments and investors depends partly on the comparative costs per kilowatt-hour of electricity generated. In some countries, nuclear power is currently competitive or even cheaper than fossil fuel alternatives. This is true for countries like Japan, France, Finland, and Canada, which have few fossil fuel resources. In all countries, nuclear power is characterized by high initial capital investments in comparison to coal- and gas-fired power plants. Note that although the current per-kilowatt-hour cost of nuclear-produced electricity in France, Japan, and Canada is low, these countries are not adding new nuclear plants at the moment. That may be largely due to the high cost of capital investment needed to build a nuclear plant.

Initial capital costs for nuclear power plants tend to run about 60–70 percent of per-kilowatt-hour electricity generating costs. In contrast, nuclear power has relatively low production costs, which include fuel, operations, and maintenance. The average cost to build a new nuclear reactor in the United States is estimated to be between US$1.5 billion and US$3 billion. In comparison, the cost of building an equivalent-size natural gas power plant is around US$450 million. In the past, nuclear electricity sales have not recovered capital costs, and in the United States these costs have consistently been underestimated by about three times the original estimate.

Another variable in nuclear power economics is construction time. For a country like the United States where no nuclear power plants have been built for decades, the construction time is a great unknown. It could take as little as five years or longer than ten years. It is difficult for an investor to commit to such uncertainty on the return on investment. Construction times tend to be much longer than for an equivalent natural gas plant. In deregulated energy markets, projects with long construction times are possible only with favourable interest rates and payback periods.

In addition to the capital and production costs, there are external costs such as those for managing and disposing of nuclear waste. The U.S. Nuclear Waste Policy Act required utility companies to charge ratepayers US0.1¢ per kilowatt-hour for a Nuclear Waste Fund, which would cover the costs of disposing of spent fuel in a geologic repository. From 1983 to 2002, U.S. ratepayers paid over US$16 billion into the Nuclear Waste Fund, but only US$8 billion has been spent on the characterization of Yucca Mountain in Nevada, because the U.S. Congress has used the rest to defer budget deficits.

Other external costs are those for decommissioning nuclear power plants and those to cover insurance funds in case of catastrophic accident. Decommissioning costs may, in fact, be underestimated. In the United States, two studies showed that estimates of decommissioning costs had risen from about US$300 million per reactor to US$500 million per reactor in one year. Because evidence suggests that these costs may continue to rise, and because no light water reactor has been completely decommissioned (including spent-fuel removal), there is reason to believe that utility companies may not be collecting enough money to cover these costs in the future.

An additional source of economic uncertainty with nuclear power is the potential for high external costs from plant aging. As a plant ages, large and expensive pieces of equipment may need to be replaced. Moreover, even if the plant owner sees no need for part replacement, the regulators may require it. An example is the experiences of the Davis-Besse plant in the United States. It was a surprise to both the plant owners and the NRC that holes developed in the reactor head, which, if they had gone completely through the head, would have resulted in a large accident. The reactor owner has been trying to fix the problem, but the reactor has been off-line for over a year now and the NRC has yet to allow it to resume operations.

A tenfold expansion in nuclear power would occur only if initial capital costs were somehow controlled. This could be done via investment guarantees and government subsidies, as some in the U.S. Senate are trying to provide. To be successful, nuclear power has to compete with other energy sources, and some renewable technologies may be less costly and more competitive in the future than they are now.

TYPES OF NUCLEAR POWER PLANT DESIGNS

Nuclear reactors harness the energy produced by the splitting of atoms (fission), usually uranium, though sometimes plutonium. In doing so, a reactor requires two processes: moderation of the energy of the neutrons that split the atoms to the energy needed to maximize fission, and cooling of the reactor components — to safely run the reactor. The coolant also performs the important job of transferring heat energy produced by the reactor to the turbines that produce electricity.

Reactors are generally distinguished by the types of fuel, moderator, and coolant that they use. In the United States, reactors were developed from those used in nuclear submarines. The most successful of these was the light water reactor, which used fuel slightly enriched in uranium-235. Light water reactors use regular water as both coolant and moderator. Two types of light water reactors are in use in the world today: pressurized water reactors, in which the core of the reactor and its water are kept under high pressure so that the water does not boil, and boiling water reactors, in which the water is allowed to boil, generating steam to run the turbines.

Britain and Canada decided to base their initial reactor designs on uranium fuel that didn't require enrichment. Britain developed gas-cooled reactors, which used carbon dioxide gas as a coolant and graphite as a moderator. These are known as Magnox reactors after the magnesium-rich alloy used to clad the uranium metal fuel. Later models of Britain's gas-cooled reactors used slightly enriched uranium dioxide fuel. Canada developed the CANDU (short for CANadian Deuterium Uranium) reactor, which uses natural uranium as a fuel and heavy (deuterium-based instead of hydrogen-based) water as both coolant and moderator.

Two other reactor designs merit mention. The Soviets designed a light water–cooled, graphite-moderated reactor known as the RMBK, the most infamous example of which is the Chernobyl reactor. Many of these reactor types are still in operation in Russia and Eastern Europe. The other important design is the liquid-metal fast-breeder reactor. This reactor uses fast neutrons and does not need to slow them down with a moderator. The coolant used is either sodium or lead-bismuth. Although a number of countries had fast-breeder reactor programs, many of them shut down due to cost and technical difficulties. Only three of these reactor types are in operation in the world now.

OBSTACLES TO NUCLEAR POWER EXPANSION: SAFETY

For those who remember them, the experiences of living through the Chernobyl and Three Mile Island nuclear power plant accidents were nail-biting moments. The 1979 Three Mile Island plant accident did not result in large releases of radiation, but the seeds of uncertainty planted in the public's mind spelled the end of expansion for the nuclear industry in the United States. The 1986 Chernobyl accident, unfortunately, was a different story; it resulted in forty-two immediate deaths (from exposed workers) and has caused a twenty-five-fold increase in childhood thyroid cancers in nearby Belarus. It will also likely result in an additional 6,500 cancer deaths in nearby residents and "liquidators" who worked to contain the radiation. A third reactor accident — involving a nuclear-weapons-production reactor, not a civilian power reactor — at the Windscale plant in Sellafield, England, in 1957 released radioactivity in amounts between that of Three Mile Island and Chernobyl.

The problem with safety issues for nuclear power is the public's well-founded fear of radiation — which can injure or kill but cannot be sensed — based on the destruction wrought by the United States' use of nuclear weapons in Japan at the end of World War II. Nonetheless, many years have passed since the nuclear industry experienced an accident. The emphasis now is on safety issues from aging nuclear power plants and the enhanced safety of new plant designs.

The existing fleet of nuclear power plants is aging fast. For example, the average age of the world's operational nuclear power reactors is twenty-one years. As mentioned earlier, the United States is already beginning to extend the licensed lifetimes of nuclear power plants from forty years to sixty years. One example of an aging problem that could affect safety is the corroded lid on the reactor at the Davis-Besse plant.

A tenfold increase in the number of reactors would greatly increase the potential for safety issues. It would certainly over-burden existing regulatory agencies, which would have to adjust accordingly, including increases in government appropriations. With so many reactors, would all equipment, operators, and regulators be of top quality?

There are a number of new reactor designs that claim to be safer to operate than the more ubiquitous light water reactor designs of the 1970s. Some of these new reactor designs have not fallen far from the original "tree" and are based on the light water reactor workhorse in use in the United States and much of Europe and Asia. One advantage of these systems is that there will be a standardized plan for the reactors. U.S. nuclear reactors have unique designs, which has not allowed the U.S. nuclear industry to take advantage of economies of scale. The new design will be simpler and the plants will have a longer life, about sixty years. The simpler design will reduce the probability of a reactor accident. Finally, the reactors will be designed to burn fuel longer, to reduce waste volumes. These designs are mostly for large-scale plants, but some are mid-size (600-megawatt) designs.

Nuclear engineers have not limited themselves to simply modifying existing reactor designs; some have designed advanced reactors. One is the high-temperature gas reactor (HTGR), the first of the new generation of which is planned for construction in South Africa. Though the design is not new, it takes advantage of a modular plan (so that additional modules can be added to a single site to increase capacity) and a more accident-resistant fuel. These reactors are expected to be much more efficient than existing nuclear power plants. The HTGR uses helium as a coolant and operates at high temperatures, about 950°C. The fuel is designed either as "pebbles" of uranium coated with carbon and silicon carbide or as uranium embedded in graphite and arranged in hexagonal prisms. The pebble type of fuel is planned for the

South African reactor, thus it is termed the "Pebble Bed Reactor."

Although HTGRs are described as "inherently safe" in design, they do have some potential safety issues. One is the fact that they are designed without containments — large, reinforced structures built around the reactor vessel itself. A containment is what the Chernobyl reactor lacked, and so when it melted down, there was no structure available to contain the radioactivity. High-temperature gas reactors can be designed without containments because the probability of a meltdown is very low, due to the fuel's ability to slow the fission process (splitting of uranium atoms) as the temperature increases. On the other hand, the reactors are susceptible to fire if air or water comes into contact with the fuel, and these reactors produce more spent fuel than do comparable-size light water reactors.

The other advanced design under discussion also is not new, but entails numerous obstacles which relegate it to the far future, if it is to be used at all. This is the fast neutron reactor, which typically takes advantage of the fast neutrons emitted by

NUCLEAR WASTE REMAINS AN UNRESOLVED PROBLEM IN ALL COUNTRIES THAT USE NUCLEAR POWER FOR ELECTRICITY PRODUCTION.

plutonium and therefore requires plutonium fuel. The idea behind these reactors is that they not only produce electricity but also replace or breed their fuel through nuclear reactions with a uranium "blanket," thus they are often referred to as fast-breeder reactors. For coolant these reactors require some type of liquid metal such as sodium or lead-bismuth; others require a gas.

Current designs of these plants are exorbitantly expensive. In addition, the use of plutonium fuel can lead to diversion for and proliferation of nuclear weapons.

OBSTACLES TO NUCLEAR POWER EXPANSION: NUCLEAR WASTE

Nuclear waste remains an unresolved problem in all countries that use nuclear power for electricity production. Considering that 20 to 30 tonnes of used nuclear fuel are produced per gigawatt per year, the current capacity of the world's nuclear power plants produces between 7,000 and 11,000 tonnes of spent fuel annually. In many countries, this spent fuel continues to reside at reactor facilities, awaiting final disposal. In some countries, such as France, the United Kingdom, Russia, Germany, and Japan, this fuel has been transported to reprocessing facilities, where its unused uranium and newly produced plutonium are extracted. Only one of these reprocessing countries so far has succeeded in using a large portion of the plutonium as fuel in existing reactors; the rest simply stockpile the separated plutonium.

The reprocessors are not off the hook in dealing with nuclear waste, however. They have reduced the overall volume of high-level waste, but they have vastly increased the volumes of low-level and intermediate-level wastes. And the high-level waste contains all the thermally and radioactively hot materials that the original spent fuel contained. When disposing of waste, the volume is not as important as the heat production.

The international consensus for solving the problem of high-level waste is to dispose of it in geologic repositories, whether in the form of spent fuel or vitrified high-level reprocessing waste. Most countries with nuclear power are developing a geologic repository, though the task has proved more difficult than previously thought. The United States, Finland, and Sweden have perhaps the most "advanced" waste disposal programs, though none have come close to actually opening a repository.

The issues facing successful disposal of nuclear waste fall into two main categories: political and technical. Though many in the nuclear industry claim that the nuclear waste problem is technically solvable, there are still many uncertainties attached to the science and engineering of nuclear waste disposal. First of all,

IT IS NOT CLEAR THAT, GIVEN THE INHERENT UNCERTAINTY IN ENSURING SAFETY WHEN DISPOSING OF NUCLEAR WASTES, IT WILL EVER BE POLITICALLY FEASIBLE TO OPEN A REPOSITORY.

it is one endeavour for which we will never know the results. If the waste leaks 1,000 years from now and affects humans living near the repository site, a result that is judged a failure by most repository regulations, we won't know it.

Second and perhaps more important is the fact that disposal is using the barriers provided by the local geology in addition to the engineered ones of the waste canisters, the tunnel, backfill, and others. Because one of the main barriers to the release of radionuclides to the environment is the geology, predictions of geologic conditions and behaviours of radionuclides and engineered materials in the geologic system over geologic time periods are essential to ensuring a successful repository location. These predictions rely on geology, a retrodictive (explaining the past), not predictive, science. Therefore, geologic disposal of nuclear waste will entail some amount of uncertainty. And this is the link to the political issues.

It is not clear that, given the inherent uncertainty in ensuring safety when disposing of nuclear wastes, it will ever be politically feasible to open a repository. There are a number of models for attempting to do so, ranging from the democratic

144

methods of France, Finland, and now Germany, where the public in the site location is given ultimate veto power over potential sites, to the more top-down approach favoured by the United States. U.S. site selection was essentially done by Congress in the Nuclear Waste Policy Act Amendments of 1987, in which it forewent the plan to characterize and select from three sites to one, the Yucca Mountain site in Nevada.

The problem for a future with a tenfold increase in nuclear power production is the huge amounts of waste produced. To manage and dispose of an annual production of 70,000 to 110,000 tonnes of spent fuel will require a new way of dealing with the waste. For comparison, the Yucca Mountain site in the United States is currently designed to hold 70,000 tonnes of waste. The tenfold increase would require tens to hundreds of Yucca Mountain–type sites. It's not clear that this will be either technically or politically feasible.

The decision may be to wait for a better alternative to geologic repositories to come around. But this would mean that an industry is allowed to continue to produce highly toxic wastes without an implemented and proven plan to deal with them. One technology on the horizon is transmutation of nuclear waste, which allows the alteration of many long-lived radionuclides into shorter-lived ones. This technology will still require some type of geologic repository for the disposal of the shorter-lived radionuclides, though. It will also require the construction of new and expensive reactors or accelerators.

OBSTACLES TO NUCLEAR POWER EXPANSION: NUCLEAR WEAPONS PROLIFERATION

The hardest part of making a nuclear weapon is obtaining the nuclear materials to power it. Therefore, perhaps the most serious problem with a large expansion of nuclear power is the threat to international security from the proliferation of nuclear weapons. There has always been a direct connection between nuclear power and nuclear weapons because the materials used to power the weapons are the same as those used to power reactors, the fissile materials plutonium and highly enriched uranium. England obtained some of the plutonium for its nuclear weapons from power reactors. India got its first significant quantities of plutonium, used for a "peaceful nuclear explosion" in 1974, from a research reactor of Canadian power reactor design.

Our international system to detect the diversion of fissile material has already been challenged by Iraqi and North Korean diversions. How would the system handle ten times more reactors? Clearly, proliferation-resistant nuclear energy technologies are required.

One potential proliferation-resistant power reactor technology is the Radkowsky concept, which uses thorium instead of uranium fuel in existing light water reactors. The advantage here is that less plutonium is produced when the fuel is irradiated or "burned" in the reactor than in typical light water reactor designs. Thorium fuel is not a complete solution, though, as it produces the isotope uranium-233, which can be used to make nuclear weapons. The idea is to dilute this with uranium-238 to levels impossible to make usable nuclear explosions.

Another suggestion is to use high-temperature gas reactors. With this type of reactor, proliferation resistance is obtained through the high burnup (long fuel irradiation time) of the fuel, which creates plutonium isotopes that make the manufacture of nuclear weapons difficult. Of course, the reactor need not be operated in that manner. Furthermore, HTGRs use fuel that is more enriched in uranium-235 compared with that used in light water reactors. Thus, countries using HTGRs may need enrichment technologies. Once uranium is enriched to 20-percent uranium-235 (the level needed for many HTGR designs), it is relatively easy to enrich further to 90 percent, posing proliferation risk.

Finally, some have suggested fast-breeder reactor technologies in which plutonium is not separated from the fission products in the fuel. The fission products make the use of this material as a nuclear weapon impossible. Of course, a country with the right resources can still separate the plutonium from the fission products. Moreover, the use of breeder reactors creates more plutonium, increasing the diversion/theft problem.

Used nuclear fuel from light water reactors cannot be used in nuclear weapons, though the plutonium in it, which makes up about one percent of the mass of the fuel, can be. Perhaps the main problem with nuclear power is that some of the associated technologies — reprocessing of spent fuel to separate plutonium and uranium enrichment technologies — pose the highest proliferation threats. Separated plutonium can easily be fashioned into nuclear weapons by knowledgeable people. Highly enriched uranium is arguably even easier to make into nuclear weapons. Thus, for years the world has safeguarded these technologies in non–nuclear weapons states.

The problem with a tenfold expansion in nuclear power is the resulting growth and spread of these technologies. Clearly, they would have to be controlled. Perhaps the best way to do so would be through the use of international uranium enrichment, reprocessing, and reactor production facilities. An additional step to ensure proliferation resistance would be internationalizing nuclear waste disposal, so that all nuclear material was controlled and accounted for from cradle to grave. The question is, how to internationalize these technologies? Who would host these facilities — and who would be dependent upon the good will of the hosts?

The problem with internationalizing these facilities is that it would create two classes of country: those deemed responsible and stable enough to host these facilities, and the rest. Furthermore, many countries find nuclear energy attractive because it allows them to become independent of others for their energy resources. Internationalizing the nuclear fuel production facilities would continue the energy dependence of many countries.

Finally, there is a problem with the large existing stocks of separated plutonium in the civilian energy sector. Over 200 tonnes of separated plutonium remains in storage in the United Kingdom, France, Russia, and other countries, with no immediate plans for its use. Though the material is (for the most part) well-guarded and accounted for, there remains the possibility of diversion or theft. These stockpiles would need to be responsibly dealt with before the step was taken of making a large expansion in nuclear power. This would increase the security of all nations by showing that there were no intentions to use this material in nuclear weapons.

OBSTACLES TO NUCLEAR POWER EXPANSION: SECURITY AGAINST THEFT AND SABOTAGE

Concerns about security and terrorism at nuclear power plants are not new, but they were certainly highlighted by the September 11, 2001, terrorist attacks in the United States. Nuclear power plants have two main vulnerabilities: the core of the reactor and its spent fuel pool storage. Depending on the reactor design, both can be vulnerable. Some reactors are designed with the spent fuel pools inside the containment; in other designs they are outside, with less protection against attack and more potential for radiation release. Some reactors are designed with spent fuel pools on floors above ground level. If the pool in such a reactor was damaged, it could potentially drain the cooling water, leading to a fire and fuel meltdown. The reason spent fuel pools are of concern is that they often contain many times the amount of fuel in the reactor core, and thus the potential for fire is higher in the event of sabotage.

Most nuclear reactors are facilities that restrict the persons allowed to enter. Nonetheless, the security threat exists. The threat could be from an outside attack from the air, water (many nuclear power plants are located on bodies of water), or ground and could be assisted by one or more insiders, helping from

THERE ARE NO OBVIOUS SOLUTIONS TO THE SECURITY PROBLEM AT THIS TIME, SHORT OF GUARDING THE FACILITIES AS NUCLEAR WEAPONS FACILITIES ARE GUARDED, A PROPOSITION FAR TOO COSTLY FOR THE NUCLEAR INDUSTRY.

within the facility itself. The problem of security at nuclear plants is exemplified by the over 50-percent failure rate of security tests done at U.S. reactors. With a tenfold increase in nuclear power plants, the risk from terrorist attack also increases.

Perhaps the solution to the risk is inherently safe reactors and inherently safe fuel. There are very few new reactor designs that have these qualities, though the Swedish-designed Process Inherent Ultimate Safe (PIUS) reactor may come close. There are no obvious solutions to the security problem at this time, short of guarding the facilities as nuclear weapons facilities are guarded, a proposition far too costly for the nuclear industry.

THE BOTTOM LINE

An expansion to more than 3,500 nuclear reactors worldwide implies first tackling the major issues of cost, safety, waste, proliferation, and security. The cost of nuclear power must be competitive not only with fossil fuels but also with renewable energy technologies, such as wind, that will become mature in the twenty-first century. Safety and cost are linked, and the new, safer reactor designs must prove to be cost competitive as well, in both the industrialized and developing worlds. The nuclear waste problem must be proven solvable, given the large amounts of waste expected to be produced with a tenfold increase in generating capacity. And perhaps most importantly, there must be enough international infrastructure and organization to deal with the proliferation and security issues.

One other problem will need to be solved before expanding nuclear power. Currently, there are no plans for phasing out nuclear power — decommissioning all the reactors, ensuring that no material is diverted to nuclear weapons or terrorists, and disposing of the wastes. Given that one day nuclear power will no longer be needed, how will phase-out occur with ten times as many reactors? Is it ethical to embark on such an increase without definite plans to deal with any of these issues? In comparison to coal-fired plants, where there is no worry about diversion of fuel to nuclear weapons, the problem of nuclear power phase-out is much more complicated.

Because of all the potential problems, many of them very serious, that a tenfold increase in nuclear power will bring, a world with increased nuclear power will not be modelled on the current situation of nuclear power use. There will have to be a way to deal with the waste problem, because it will not be solved with geologic repositories — the public simply would not allow that (seeing how they have yet to allow it). If that is the case, then some type of reprocessing and/or transmutation will be required to deal with the spent fuel. And in that case, there will have to be tight controls to avoid the proliferation of nuclear weapons. Because the amount of plutonium produced in such a scenario is on the order of 150,000 bombs' worth per year, reprocessing will need to be done in a single international centre. I suggest a single centre because the non-proliferation advantage would be lost to competition among fuel suppliers if there was more than one centre. And one centre would be the most equitable, if it were truly international.

In the end, such a large expansion in nuclear energy may only be possible if the connection between nuclear energy and nuclear weapons is fully broken. For this to be done, nuclear weapons would likely have to be abolished and sworn off by all countries. In today's world, it is difficult to imagine such a scenario.

At this time, it is not at all clear that these issues will be resolved in the long term. If they are not, the potential for harm from reactor accidents, nuclear weapons use, or terrorist attacks will make a large expansion of this energy form undesirable, even given its beneficial effects on the climate. In the end, with nuclear power we must weigh the risks and their potentials: could there be a nuclear bomb from diverted nuclear materials, and is the explosion of one of these weapons worth the money saved from renewable energy technologies and the advantages for the climate? I'm not sure we are able to answer that question right now.

A HISTORY OF ENERGY (CONTINUED)

1877 In the first public use of electricity in Canada, the Montreal Harbour Commission uses electric light to illuminate its waterfront.

1879 Thomas Alva Edison discovers that a small filament of heavy paper in a vacuum, when exposed to electricity, can sustain a long-lasting source of light, and on New Year's Eve he demonstrates the incandescent light bulb at Menlo Park, New York. Although Edison does not technically invent the light bulb — that honour is generally credited to the English chemist Humphrey Davy in 1811 — Edison's discovery allows for the light bulb's practical construction from available material for home use. Before it can be commercially available for that purpose, however, Edison is required to invent seven other system elements, including the parallel circuit and light sockets with on/off switches. These inventions will lead directly to the development of the modern electric industry. In fact, his original company, Edison General Electric, is the precursor to GE. Siemens & Halske build the world's first operating electric train for Bush Mills, Northern Ireland.

1880s The Swedish-born engineer John Ericsson, who moved to the United States to help the Union forces in the design of a new kind of warship, develops solar-driven engines for ships. Charles Fritts constructs the first solar cell using the light-sensitive metal selenium; between 1 and 2 percent of the light received by the cell generates electricity.

1881 While strolling through Budapest's Varosliget city park, the physicist Nikola Tesla discovers the principle of the rotating magnetic field. He imagines an iron rotor spinning in a magnetic field, produced by the interaction of two alternating currents. This will lead to the development of the induction motor, a first step towards the development of alternating current (AC), which can travel greater distances than direct current (DC). Tesla will patent a device to induce electrical current in a piece of iron spinning between two electric coils. A few years later George Westinghouse, head of the Westinghouse Electric Company in Pittsburgh, will buy Tesla's patents, paving the way for induction motors that run household appliances, industrial motors, as well as the world's first large-scale hydroelectric plant at Niagara Falls, which will bear Telsa's name.

▼ CONTINUED ON PAGE 178

07.

THE PERFECT ENERGY: FROM EARTH, WIND, OR FIRE?

PETER FAIRLEY

EDITORS' INTRODUCTION

Partly because of the potential hazards of nuclear power, there is growing interest in other sources of energy that might be able to provide large-scale, clean power. The next writer, Peter Fairley, has spent much of his career writing for numerous magazines and publications about the technologies that might provide a cleaner alternative for fueling our future. Specifically, he has looked at three technologies: wind power, solar power, and clean coal to see to what extent they might be part of a cleaner energy future.

There has been growing attention paid to these potential clean technologies, particularly wind power, which has been the fastest-growing new power source for many years now on the back of technological innovations that have helped bring the price of wind power down. Windmill technology has changed dramatically, from the four-bladed post mills to the fabled Dutch tower mill, to the millions of steel-bladed fan-type windmills used in the United States to pump water during the nineteenth and early twentieth centuries, to today's jumbo-sized turbines.

It wasn't until 1888 that a man named Charles Brush in Cleveland, Ohio, built the first windmill to generate electricity. Seventeen metres in diameter, with a gearbox that turned a direct generator, his windmill could generate about 12 kilowatts of power, which is to say, not too much. It was the Russians who, in the 1930s, began to generate utility-scale power from wind, with their

Balaclava wind generator, capable of producing 100 kilowatts.

Efforts to strive after wind have continued, spurred on in North America by the oil crisis of 1973–74. And in the past decade, new, lighter and stronger composite materials and computer technology have allowed wind power to become a major energy player. In our home city of Toronto, a large wind turbine was recently installed on the Canadian National Exhibition grounds where the Gardiner Expressway cuts by Lake Ontario. Although there were undoubtedly better locations from an energy standpoint, there were none from a PR standpoint: millions of people who commute past now have daily cause to consider the potential of wind energy. The sleek, efficient European design stands 94 metres tall, higher than the Royal York Hotel downtown, and the 29-metre blades will provide enough energy for 250 homes.

The windmill is a symbol of a new power industry, one where technological advances are pushing down the costs of less-polluting sources of energy. Peter Fairley has spent much of his career writing about energy technologies for magazines such as MIT's *Technology Review* and *Canadian Business*, and we asked him to investigate whether technological advances will usher in a new era of clean power or whether they are merely striving after wind.

ADVOCATES OF RENEWABLE ENERGY are piercing the last nagging doubts about the viability of their technologies. It is now accepted wisdom that they have an important role to play in powering our economies. Consider the new image of wind power. Even officials in the Bush administration, its rise heavily financed by the oil and gas industry, can't help but respect the emerging wind industry in the United States. "Today we're celebrating moving a mature, renewable technology from the lab to the marketplace," said Spencer Abraham, George Bush's Secretary of Energy, in 2001 as he ramped up wind power purchases by the U.S. federal government.[1] People like Abraham have to say nice things about wind power today because it is the fastest growing of all energy sources. In 2001 alone, wind developers in the U.S. installed 1,700 megawatts' worth of new wind turbines — enough to power approximately 400,000 homes — nearly doubling the nation's wind-power capacity. In Canada, wind power grew a remarkable 47 percent in 2001, reaching 215 megawatts. World-wide, developers and utilities installed a record 6,868 megawatts of new wind-power capacity in 2002, boosting total capacity by 28 percent.[2]

The number of installations of electricity-generating solar panels is tracing an equally promising trajectory, doubling every three years. Rooftop solar systems are paying their way right now in places like California where conventional energy is expensive and governments eager to cut pollution and ease congestion on power lines help cover their cost. More than 200,000 households in the U.S. derive at least some of their power from solar cells, including the White House, which quietly commissioned its first *photovoltaic* panels in January 2003. The 9-kilowatt, 167-panel system marks solar power's return to the citadel of global power after a twenty-year hiatus. (Ronald Reagan tore down the solar water heater that Jimmy Carter installed on the White House roof.)

The promising signs of health in the wind and solar industries owe much to advanced materials, innovative designs, and other technological advances that have delivered cheaper and more reliable renewable devices to harvest the energy of the wind and sun. However, the gains achieved to date by renewable energy technologies are easily overstated. Wind power still produces less than one percent of electricity in North America, and solar, despite nearly quadrupling in sales over the last five years, still accounts for only 0.04 percent of worldwide power generation. The consensus in the U.S. solar-power industry is that continued installation of solar panels at today's impressive rates for another three decades will lift solar's contribution to just 10 percent of peak power generation in the U.S.[3] For clean power to truly power our economies, its economics must become even more attractive, and that will require more than technological change. Inherently cheaper technologies for harvesting the energy of the wind and the sun already exist, and there are troubling signs that today's flourishing early markets for renewable energy are stifling their application. Limited subsidies for wind and solar power are sparking investment, but the market players making the investment decisions are pursuing a low-risk approach designed to maximize profitability today rather than revolutionize energy markets.

To transform our climate-altering energy habits, we must not only make the economics of renewable energy *more* attractive. Canadian and U.S. governments must make the economics of dirty power *less* attractive, by billing conventional power producers for the pollution they create. Billing utilities in the U.S. and Canada for the health and environmental impacts associated with burning coal, for example, would more than double the cost of

MANUFACTURERS HAVE REDUCED THE COST OF WIND TURBINES MORE THAN FOURFOLD SINCE 1980

coal-fired power.[4] That would do more than reward wind developers and solar users who generate pollution-free power. It would unleash innovative technologies that could capture the greenhouse gases from fossil fuel power plants — technologies that could clean up even coal, but which languish due to a lack of financial incentives for responsible behaviour by the power industry.

WIND POWER'S WEAK KNEES

In just twenty years wind power has evolved from an industry marked by widespread disaster and ridicule to one that is highly successful. Manufacturers have reduced the cost of wind turbines more than fourfold since 1980, and wind farms are sprouting up in the hundreds, not just in Germany and Denmark but also in southern Alberta and west Texas. Even investor extraordinaire Warren Buffet is getting into the act. Buffet-controlled MidAmerican Energy Co., the largest utility in Iowa, plans to assemble as many as 200 gargantuan wind turbines to create what would be the largest wind farm in North America. Wind turbines are even reliable and efficient enough to be built offshore, where winds are both more powerful and more consistent. An 80-turbine, $330-million wind farm installed in the summer of 2002 off the Danish coast is the world's largest, and developers are beginning to colonize German, Dutch, and British waters, too. In North America, speculators envision massive wind farms off the coast of the Queen Charlottes and Vancouver Island, Long Island, and Nantucket. The fact that these proposals have engendered local opposition comparable to the "not-in-my-backyard" protests spurred by coal- and gas-fired power plants is a clear sign that wind power has arrived.

But while the wind industry has gained legitimacy, it has lost much of its scope for innovation. Where wind turbines of all shapes and sizes once graced wind farms, today there is a striking uniformity to their design: regardless of manufacturer, every turbine sports three stiff blades facing into the wind. It's not a particularly cheap design; on average, a modern wind farm costs about $1.5 million per megawatt to construct, compared to about $900,000 per megawatt for a conventional gas-fired power plant. In the economic calculations of power companies, the fact that the wind itself is free and non-polluting doesn't close this gap. In short, the price of building wind power must come down if it's ever to be more than a mere niche technology.[5]

To my knowledge one of the only companies offering an alternative wind turbine design that could deliver substantial cost reductions is a Bellevue, Washington–based outfit called The Wind Turbine Company. Its turbines sport only two blades but otherwise look much like any other apparatus for capturing energy from wind, until you look closely when the wind rushes by. In a gust, the blades bend back on a set of hinges. This flexibility should enable the use of machines that are 40 percent lighter than today's industry standard, and that lighter weight could mean machines that are 20 to 25 percent cheaper than today's large turbines.

Unfortunately, as we shall see, The Wind Turbine Company is finding life on renewable energy's cutting edge very lonely.

Wind turbines are like giant fans running in reverse. Instead of pushing the air with motor-driven blades, turbines use airfoils to catch the wind and crank a generator that pumps out electricity. Many of today's turbines are mammoth machines with rotors that span 80 metres. That's 20 metres wider than the wingspan of a Boeing 747. And therein lays the technology challenge. The enormous size is a necessity if commercial wind turbines are to compete economically, because power production rises exponentially with blade length. At the same time, these vast structures must be rugged enough to endure gales and extreme turbulence.

In the 1970s and '80s, U.S. wind-energy pioneers made the first serious efforts at fighting these forces with lightweight, flexible machines. Several startups installed thousands of such wind turbines. Most were literally torn apart or disabled by gusts. Taking lightweight experimentation to the extreme, General Electric and Boeing built much larger prototypes — behemoths with 80-, 90-, and even 100-metre-long blades. These also proved prone to breakdown. In some cases their blades bent back and actually struck the towers.

All told, U.S. companies and the U.S. Department of Energy spent hundreds of millions of dollars on these failed experiments in the 1980s and early 1990s. "The American model has always gravitated toward the light and the sophisticated and things that didn't work," says James Manwell, a mechanical engineer who leads the University of Massachusetts's renewable-energy research laboratory in Amherst, MA.[6]

Into these technology doldrums sailed researchers from Denmark's Risø National Laboratory and Danish companies like Vestas Wind Systems. During the past two decades they perfected a heavy-duty version of the wind turbine — and it has become the

Microsoft Windows of the wind-power industry. Today, this Danish design accounts for virtually all of the electricity generated by wind worldwide. These sturdy Danish designs had little of the aerodynamic flash of the earlier U.S. wind turbines. The Danish designs were simply braced against the wind with heavier, thicker steel and composite materials. They were tough — and they worked.

What's more, in recent years power electronics — digital silicon switches that massage the flow of electricity from the machine — further improved the basic design. Previously, the turbine's rotor was held to a constant rate of rotation so its alternating-current output would be in sync with the power grid. The new devices maintain the synchronization while allowing the rotor to freely speed up and slow down with the wind. "If you get a gust, the rotor can accelerate instead of just sitting there and receiving the brute force of the wind," says Manwell.

Mastering such strains enabled the Danish design to grow larger and larger. Whereas in the early 1980s a typical commercial machine had a blade span of 12.5 metres and could produce 50 kilowatts of power — enough for about a dozen homes — today's biggest blades stretch 80 metres and crank out 2 megawatts. A single machine can now power more than 500 homes.

The newest challenge facing the Danish design is finding ways for it to withstand the corrosive and punishing offshore environment, where months can pass before a mechanic can safely board and fix a turbine. Vestas, for one, is equipping its turbines with sensors on each of their components to detect wear and tear, and backup systems to take over in the case of, say, a failure in the power electronics. Its approach is being tested in an 80-turbine Danish wind farm, situated in shallow water 14 to 20 kilometres off the Danish coastline, which meets 2 percent of Denmark's power needs.

These upgrades make big, heavy turbines more reliable,

but they don't add up to a fundamental shift in the economics of wind power. Nations like Denmark and Germany are prepared to pay for wind power partly because fossil fuels are so much more costly in Europe, where higher taxes cover environmental and health costs associated with burning them. (About 20 percent of Denmark's power comes from wind.) But for wind power to be cost competitive with fossil fuels in North America today, the technology must change. "To do that you've got to take some risks," says Bob Thresher, director of the National Wind Technology Center at Rocky Flats, Colorado, the U.S. Department of Energy's proving ground for wind-power technologies.

Nobody is taking greater risks than The Wind Turbine Company. What marks its design as such a departure from the Danish model is not only its hinged blades but also their downwind orientation. The Danish design faces the blades into the wind and makes the blades heavy so they won't bend back and slam into the tower. The Wind Turbine Company's design can't face the wind — the hinged blades would hit the tower — so the rotor is positioned downwind instead. In normal conditions the blades spin freely, the entire turbine swivels according to wind direction, and a gearbox amplifies blade rotation speed so a generator can produce power. In high wind or erratic conditions, hydraulic dampers allow blades to flex up to 15 degrees downwind and 5 degrees upwind to shed excess wind force. Finally, as mentioned earlier, the turbine uses two blades, rather than the three in the traditional design, in order to further reduce weight.

Advances in the computer modelling of such dangerous forces as runaway vibrations (not unlike the feedback from a microphone) helped the design's development. Flexible blades add an extra dimension to the machine's motion; so does the fact that the whole machine can swivel freely with the wind. (Traditional designs are driven to face the wind, then locked in place.) Predicting, detecting, and preventing disasters — like rapidly shifting winds that swing a rotor upwind and send its flexible blades into the tower — are control challenges even with the best design. "If you don't get that right, the machine can literally beat itself to death," says Ken Deering, the company's vice president of engineering.

There is a lot of experience with such failure. In fact, earlier efforts at lighter designs were universal failures — disabled or destroyed, some within weeks, by the wind itself. When The Wind Turbine Company's prototype was erected at Rocky Flats in 2000, there were worries that this machine, too, would beat itself to death. Thresher says some of his staff feared that the machine, like its 1980s predecessors, would not long escape the scrap heap. Today, despite some minor setbacks, those doubts are fading.

The Rocky Flats prototype is the best hope in years for a lightweight design that will finally succeed. "I can say pretty unequivocally that this is a dramatic step in lightweight [wind turbine] technology," says Thresher. "Nobody else has built a machine that flexible and made it work."

If the design works in mass-produced machines, it will change the economic equation for wind power. "The question would become, 'How do you get the transmission capacity built fast enough to keep up with growth?'" says Ward Marshall, president of the American Wind Energy Association, a Washington, D.C.–based trade group. "You'd have plenty of folks willing to sign up."

Nevertheless, The Wind Turbine Company remains out in the cold — unable to raise the funds to kick-start mass production. The problem is financing. U.S. banks won't touch wind farms, and the European banks that do are skittish about technological innovation. It didn't help that *Windpower Monthly*, the industry bible, featured the company's downwind technology under the headline, "*Radical* design goes commercial" (emphasis added).

Marshall, a former wind-power developer at U.S. power giant American Electric Power and now a sales manager for turbine producer GE Wind Energy, voices his own personal concerns about The Wind Turbine Company design. "I'm never going to say it's impossible, but it's just extremely difficult to control those dynamic loads and model them correctly. I still think there's some significant technology risk in heading in that direction," says Marshall. But he says it's the banks' opinion that really matters. "Ultimately the end of the road is the bank that's going to put the money forth. Those folks are getting pretty darn comfortable with the Vestas type concept," says Marshall. That's fine if you want to sell a few thousand wind turbines a year to take advantage of today's limited government supports for wind power. It's downright disastrous if you want to sell cheaper machines in the tens or hundreds of thousands, machines that could revolutionize the way power is produced in North America.

Ken Deering notes the irony in what a thoroughly unradical industry the wind business has become. His frustration is palpable: "We once harboured the view, maybe it was fantasy, that if we could make a reasonable case for a significant advance, people would be knocking our doors down. So far the door has not been knocked down."

SELLING SOLAR SHORT

The advantages of solar power are even clearer than wind power's. Every minute, the sun pounds the surface of the Earth with more energy than the entire world consumes in a year — a potential source of virtually unlimited, clean, and free electricity. And, unlike wind power, solar power can be harvested by solid-state devices, cousins of the transistors that do the thinking in your cell phone or laptop. Once installed, solar panels operate for decades without oil changes or gear adjustments.

Until recently the high cost of the materials used in solar cells relegated the technology to powering satellites, high-tech backwoods cabins, and communications towers beyond the reach of power lines. Conventional solar cells cost so much because most are made from the same relatively expensive silicon semiconductors used in computer microchips. Recently, the cost of silicon has dropped and manufacturers have found ways of slicing it thinner. Hence consumption is increasing. Newer technologies to spread silicon, or other semiconductors, in thin films could cut the cost of solar power even further, delivering drastic cost reductions to enable solar cells to effectively compete with coal-, oil-, and natural-gas-generated electricity across the globe.[7] But while there have been great strides made in the laboratory, translating these into new products has proven tough, and even some large companies with deep pockets are abandoning the effort. Just as The Wind Turbine Company's hinged-blade turbines are finding it difficult to make the transition from engineering curiosities to mass-produced products, a short-term rush for profits is holding back promising solar technologies.

Solar cells or photovoltaics exploit the same property that makes semiconductors so vital to the computer industry: their

BP IS NOT ONLY ONE OF THE WORLD'S LARGEST OIL COMPANIES. THROUGH ITS SUBSIDIARY, BP SOLAR, IT IS ALSO ONE OF THE TOP THREE PRODUCERS OF PHOTOVOLTAIC PANELS.

ability to guide the flow of electrons. When sunlight strikes the surface of a semiconductor, the photons of light transfer their energy to electrons in the material. A built-in electric field draws the excited electrons out of the top of the panel, creating a current of electricity. Most of the solar panels sold today, like most of the computer chips, employ crystals of silicon that are expensive to manufacture, pricing solar power out of a market dominated by artificially cheap fossil fuels. "If you want to compare purely on a dollar-per-watt basis, solar power is three to four times more expensive right now," says Atul Arya, vice president for group technology strategy at BP.

BP is not only one of the world's largest oil companies. Through its subsidiary, BP Solar, it is also one of the top three producers of photovoltaic panels. Since the early 1980s the search for low-cost technology has focused companies like BP Solar and its competitors on developing photovoltaics employing cheap amorphous silicon, and even semiconductor alloys, that can be quickly spread into a thin film just a few thousandths of a millimetre thick — about a hundred times thinner than the silicon crystals used in conventional solar cells. Because these thinner solar cells require less semiconductor material and are amenable to mass production, they were projected to be significantly cheaper.

Ken Zweibel, who leads thin-film development at the U.S. Department of Energy's National Center for Photovoltaics in Golden, Colorado, where many of the technology advances are being made, predicts that thin films will deliver highly efficient solar cells at one-quarter to one-fifth the cost of today's cells. Cheap, thin films of amorphous silicon or alloy that can capture as much as 20 percent of the sun's energy (researchers can now make films in the lab with 18-percent efficiency) could make solar cells practical for homeowners, not just in sunny California, where clogged power lines deliver the country's most expensive electricity, but in Toronto, Chattanooga, and Tampa Bay.

The low cost and inherent flexibility of these thin solar cells also means they can be easily applied as coatings on a range of materials, including glass and roofing tiles. To demonstrate this aspect of the technology, BP has installed translucent awnings embedded with thin-film photovoltaics at 250 of its gas stations around the world, keeping customers dry while powering the pumps. "It becomes a window pane, or it could be a shading application, or it could be a skylight application. We are doing all of those," says BP's Arya.

At least, that's what Arya told me in 2001 when he was chief operating officer with Linthicum, Maryland–based BP Solar and was being interviewed for a story on solar technology for *Technology Review* magazine. A year later, BP Solar abruptly changed its tune on thin films, dropping this technology of the future to invest instead in expanded production of conventional silicon-crystal photovoltaics. BP Solar abruptly ceased production at its thin-film plants in December of 2002 and, overnight, erased all mention of thin-film technology from its web pages. The corporate history page that had proudly chronicled two decades of research and development vanished.[8]

BP's withdrawal was a devastating blow to the emerging thin-films industry. Even officials at the National Center for Photovoltaics, which has been spending US$50 million each year helping manufacturers including BP Solar and the Royal Dutch/Shell Group subsidiary Shell Solar to scale up production of thin-film modules, acknowledged the damage. "You can't say this is good news. This is a blow to thin films," said director Larry Kazmerski, speaking from his office in Golden, Colorado, just a short drive from Rocky Flats, where The Wind Turbine Company's prototype bends in the breeze.

BP Solar made a business decision. Its parent, BP, promised in 1998 to derive US$1 billion in revenue from photovoltaics by 2007, and BP Solar realized that investing more in thin films was not the best way to get there. "While [thin-film] technology continues to show promise, lack of demand for the material and present economics do not allow for continued investment," said the company's president and CEO, Harry Shimp.[9] Instead, BP would make more crystalline silicon modules, which accounted for about 90 percent of its estimated US$300 million in sales in 2002.

The problem with thin films is that in today's market they are a second-class product. They are not yet much cheaper per kilowatt than crystalline solar cells, which have come down in price thanks to large-scale robotic production, an order-of-magnitude decrease in the price of silicon ingots over the past fifteen years, and sawing methods that generate thinner wafers. And thin-film solar cells don't put out as much power per panel as crystalline cells do. That means that a given roof will produce less power if it's covered with thin-film panels. Solar-power experts say that's a real drawback for buyers who are installing solar power to make a statement about renewable energy.

Thin films suffer with such buyers for another reason. They also want their solar power to show. They want a panel that looks like a crystalline panel, a panel that screams "renewable power." They want the classic metallic polka-dot motif of conventional crystalline silicon panels. BP Solar invested in architectural applications of its Powerview thin-film panels, which offer a chic black look, but architects don't want chic. Shimp explains: "When we first started talking to building materials folks and architects they said yeah, that would be great. We won invention awards for the Powerview module. But what happened was that when building owners really had to make a decision, if they were going to spend the extra amount of money to put photovoltaics in the building, they wanted it to look like photovoltaics. They didn't want it to look like an advanced material. So the aesthetics were so good, people didn't want it. If you're going to spend $200,000 for a Rolls Royce, you want it to look like a Rolls Royce. That's part of the reason that you're spending that money. They want the polka dots."

Another problem was the fact that some of BP Solar's thin films contained a semiconductor alloy with a small quantity of cadmium, a toxic heavy metal. Never mind that the panel contained less cadmium than a rechargeable battery, or that the cadmium was bound up in the panel and may have posed no threat to the environment. Like the panel's chic, unconventional look, the mere presence of cadmium worked against the product's green image. "It doesn't matter how we treat it, it doesn't matter how good the chemistry is, how safe it is, you can't get around the fact that its cadmium," says Shimp.

Shimp says that BP will continue researching thin films, and he expects that they will eventually have their day in the sun. In fact, he says the future of the solar industry is riding on them: "I think the whole industry will be coming back to thin films at some point." The reason is that the price of crystalline silicon will eventually reach a floor, a bottom line beyond which Shimp says further cost reductions will not be possible: "The only way you can break the code on direct manufacturing costs, which is the cost of materials, the cost of utilities, labour, is to lower your material costs. And the key to doing that is thin films."

So even BP Solar's CEO agrees that the future of solar is ultimately in the hands of the thin films. But, at least for the time being, Shimp is not going to invest BP's manufacturing and marketing dollars to make that future a reality.

FOSSIL FUELS HAVE IT WORSE

If you think the markets are hard on innovators in renewable energy, consider the challenge faced by innovators in the fossil fuel sector. Their most impressive accomplishment of the last century — the development of clean and efficient "combined-cycle" gas turbines — is quite literally running out of fuel. Meanwhile, the most innovative technology for cleaning up fossil fuel power plants — sequestration methods to trap the greenhouse gases they generate — can't even get a start in the market. Why? Because there is no economic incentive for power producers to use the technology. As American Federation of Scientists president Henry Kelly puts it, "Sequestration is one of the few energy efficiency technologies where there really is no economic incentive."

There has been a lot of hype about "decarbonization" — the fact that energy production has been shifting steadily from fuels that are carbon-intensive, such as wood and coal, to fuels that are less carbon-intensive, such as oil and, most recently, natural gas. Technology has been a key contributor in the latest stage of decarbonization, as advanced high-temperature steel alloys and innovative turbine designs have delivered combined-cycle turbines that generate power and heat from natural gas without the heavy soot and acid pollution associated with coal-fired power plants. Unfortunately, this latest stage of decarbonization has hit a snag that turbine technology can not address: plummeting reserves of natural gas. Rising gas prices and uncertain oil supplies are driving power producers all the way back to coal. In fact, the International Energy Agency estimates that more than 600 gigawatts of new coal capacity will be brought on line worldwide by 2020.[10] If that new capacity is delivered with conventional coal-fired power plants, those plants will pour an average of one billion tonnes of extra carbon dioxide into the atmosphere each year for their sixty to eighty years of operation. That's four times more than the annual greenhouse-gas *reduction* that Canada is attempting to make under the Kyoto Protocol.

The good news is that there is a solution: a design for a coal-fired power plant that releases as little soot and acid as a gas-fired plant and, in a trick that those efficient natural gas plants can't match, simultaneously traps its carbon dioxide.

In the lexicon of the twenty-first century, this next-big-thing in energy is called "clean coal." What sounds like an oxymoron is really the combination of two technologies: coal gasification and carbon sequestration. Gasification is how the soot and acid that long were coal's calling cards are eliminated. The coal is steam-heated to a blistering 2,700°C or hotter, forming a mixture of carbon monoxide and hydrogen gas that can be ignited to generate clean power. If desired, some or all of the hydrogen can be recovered for use as a fuel (think fuel-cell cars running on coal-derived hydrogen). Carbon sequestration is how the carbon dioxide is captured. Gasification plants produce a concentrated stream of carbon dioxide, which can be piped to an underground reservoir. Pick the right reservoir and the carbon dioxide will become trapped in the minerals and remain underground for hundreds of years.

The bad news is that in February 2003 President Bush announced that the United States would sponsor a US$1-billion, ten-year demonstration project to create the world's first coal-based, zero-emissions power plant to produce electricity and hydrogen. Spencer Abraham explained that the program, FutureGen, would establish the "technical and economic feasibility" of producing electricity and hydrogen from coal while capturing and sequestering the carbon dioxide generated in the process. Abraham called FutureGen "one of the boldest steps our Nation has taken toward a pollution-free energy future." The Canadian Clean Power Coalition, a group of utilities representing over 90 percent of Canada's coal-fired power generation, has its own ten-year plan to build a clean-coal plant. There's just one problem with all this hype: gasification and sequestration are conventional, off-the-shelf technologies that should be in use today.

James Childress, executive director of the Gasification Technologies Council, an industry group in Arlington, Virginia, says petroleum refiners already employ gasification technology to generate power and hydrogen from the nastiest low-value leftovers from the refining process. General Electric proudly states in technical presentations that it has close to 400,000 hours of experience operating turbines in gasification plants for refining giants such as Exxon, Texaco, and Shell. Gasification plants are also a

171

popular item in the chemical industry, which uses them to wring power from hazardous by-products. These are not high-tech demonstration plants. According to energy experts at Princeton University who have been studying the economics of gasification power plants, China's chemical producers have over twenty gasification plants in operation, under construction, or on order.[11]

Sequestration technology is only slightly less mature. Oil producers routinely pump carbon dioxide into slack oil wells in a process known as enhanced oil recovery. Carbon dioxide is released from a natural deposit, piped to a maturing oilfield where the bubbling crude has slowed to a crawl, and forced into the ground to literally push the remaining oil out. In the view of Howard Herzog, director of the carbon sequestration initiative at the Massachusetts Institute of Technology in Cambridge, Massachusetts, "The technology to inject carbon dioxide into the ground is well established." According to Herzog, enhanced oil recovery is a mature technology and roughly 43 million tonnes of carbon dioxide were being injected every day at sixty-seven commercial projects in the U.S. *five years ago.*[12]

Burying the carbon dioxide pollution from a power plant is somewhat less conventional, but by no means without precedent. For twenty years a plastics plant in Joffre, Alberta, has been pumping its carbon dioxide into nearby oil reservoirs, helping to yield 3.5 million barrels of oil from a previously abandoned field. In the process, the chemical plant has sequestered over 800,000 tonnes of its carbon dioxide. A more recent demonstration spans the U.S.–Canada border. Since 2000, Dakota Gasification Company — which operates a gasification plant in Beulah, North Dakota, that employs some extra chemistry to turn gasified coal into synthetic natural gas — has been shipping half of its carbon dioxide north through a 300-kilometre pipeline to an oilfield in Weyburn, Saskatchewan. Scientists from eighteen countries are watching the ground in Weyburn to verify that the buried carbon dioxide stays put. If it does, the project will tuck away as much as 20 million tonnes of carbon dioxide over the next twenty years.[13] That's about 0.3 percent of the world's total annual emissions of carbon dioxide.

So why aren't clean coal plants that marry gasification and sequestration springing up all over North America? The word from Washington is that technology advances are needed to lower the cost of gasification and sequestration. The experts sing a different tune. Consider a recent study from two of the U.S. Department of Energy's own national labs.[14] The researchers considered the role that gasification plants could play in solving California's recent energy crisis and estimated that coal gasification plants could turn a profit in California by selling their carbon dioxide for enhanced oil recovery, because there is an unmet need for carbon dioxide in the state's aging oil industry as well as strong demand for locally produced power. The problem, they report, is that building a "state-of-the-art" natural-gas-fired power plant and simply releasing the carbon dioxide pollution is even more profitable.

David Hawkins, director for climate change at the Natural Resources Defense Council in Washington, D.C., singled out this

WHAT IS NEEDED TO TIP THE SCALE IN FAVOUR OF CLEAN COAL IS TO BILL POWER PRODUCERS FOR THEIR CARBON EMISSIONS — CREATING AN INCENTIVE FOR THEM TO PUT A CAP ON THEIR STACK.

Department of Energy study in testimony before a U.S. Senate committee considering whether to regulate greenhouse gas emissions: "The good news is that DOE concludes these projects could use commercially proven technology and make a profit without any government subsidies. The bad news, according to the same study, is that as long as carbon dioxide can still be released without penalty, a project developer can make more profit building a conventional natural gas plant and venting the carbon dioxide to the atmosphere."[15] In other words, since U.S. and Canadian power producers currently pay nothing to spew carbon dioxide into the atmosphere, they have no incentive to pay the higher cost of being green.

What is needed to tip the scale in favour of clean coal is to bill power producers for their carbon emissions — creating an incentive for them to put a cap on their stack. Mandatory carbon trading programs proposed in the U.S. and Canada to meet the pollution reductions promised under the Kyoto Protocol — which both countries signed and Canada recently ratified — would do just that. The irony, of course, is that the companies that mine coal and the utilities that burn it are among the most virulent opponents of regulating carbon pollution, including carbon trading. As a result, the coal lobby is helping to ensure that gasification remains a research project instead of a commercial reality. It's a frustrating state of affairs for technology developers, says Childress at the Gasification Technologies Council: "Speaking frankly, the more stringent the requirements, the better the world it is for gasification."

ECO-PRICING

As these snapshots of technology development in wind power, photovoltaics, and fossil fuel power production show, we *can* develop the technologies needed to revolutionize our use of energy. In fact, innovative technologies that could alter our climatic future are already begging for commercialization. To freeze our impact on the Earth's climate by 2050 while meeting increasing demand for energy, we must begin installing them at an amazing rate.[16] That will happen only if these technologies are valued for their environmental contributions and thereby find the financial backing they need to become mass-produced products.

Corporate and government support to push clean-energy technologies into the market is clearly insufficient. Whether it's banks investing in wind power or industry giants developing photovoltaics, corporations generally choose short-term profitability over the risky long-term investments needed to alter an entire market. And while government officials may have the long-term vision, they lack the financial power or the operational "know-how" to effect change in the markets. Kazmerski, Thresher, and other government officials charged with advancing clean technologies would like to help push the most promising inventions into the market, but under today's "free-market" ideology it is a political no-no for the government to influence the success of particular players in an industry. Thresher acknowledges that this precludes his operation from nudging The Wind Turbine Company into the market. "It's hard for the federal government to support that stuff because that's like commercialization. That's the role of the private sector," says Thresher.

In fact, people like Thresher are lucky simply to maintain their existing capabilities to support basic research. President Bush proposed in his first budget to slash federal research on wind, solar, and most other renewable energy technologies. He lost that fight in Congress, which restored the funds. Now Bush is at it again in 2003, proposing to boost R&D investments by the Department of Energy 5.7 percent to US$8.5 billion but devoting nearly all of the increase to defence projects while freezing energy R&D. Solar and other renewable energy R&D would in fact drop in order to free up cash for priority areas such as nuclear energy.[17]

Government support for renewable energy markets is no more certain. The U.S. provides a 1.5-cent tax credit for every kilowatt-hour of wind energy produced, which in essence provides the profit margin for most wind developers in the U.S. Unfortunately the credit ebbs and flows about as predictably as the wind itself, wreaking havoc with the market for wind turbines. The U.S. wind industry raced to plug in its turbines before the credits expired at the end of 2001, and then went dormant for the first three months of 2002 until the U.S. Congress renewed them once again. As a result, just 410 megawatts worth of wind power went up in 2002 — barely 25 percent of 2001's crop. Thresher says the U.S. wind industry would probably be twice its size were it not for the on-again-off-again credits.

In Canada the federal government was late to the table with wind incentives, providing its first credits for wind-power production only in 2002. But at least it appears to have learned from the U.S. experience, locking in the incentives for five years. Canada's small solar-power industry has been less fortunate, booming and busting with the passage of one government program after another. According to industry veterans, Prime Minister Jean Chrétien himself brought the industry to its knees twenty years ago when he took over the energy portfolio under Prime Minister Pierre Trudeau and froze spending on solar demo programs for nine months, leaving the industry's fledgling players holding huge inventories. Bob Swartman, founder of London-based solar heating equipment maker Solcan, says it took his firm two years to pay off debts incurred thanks to Chrétien's move. Swartman says one competitor, Solartech, never recovered. And after a decade as Prime Minister, Chrétien has yet to install solar power at 24 Sussex Drive.

But what Swartman and other clean-energy advocates want more than government support is a level playing field for non-polluting energy technologies. In other words, they want the price of fossil-fueled power to reflect its environmental cost. A report issued this winter by Alberta's Pembina Institute for Appropriate Development cited the lack of such "eco-pricing" on polluting forms of energy production as the most significant barrier facing renewable energy in Canada: "There are no well-established markets for greenhouse gas emissions . . . as of yet and thus no financial cost for their production. Without price signals for such 'externalities,' energy projects that impact on health, society and the environment . . . are subsidized by the public through a government liability to deal with the environmental impacts in the future. To add to that, the lack of price signals means that [renewable] projects are not financially rewarded for their environmental benefits."[18] This winter Chrétien took an important step towards levelling the playing field for clean energy, a step that his counterparts in the White House have neglected to take: he instructed his government to ratify the Kyoto Protocol.

The job of implementation will fall to Chrétien's successors. If they follow through with measures such as carbon trading that raise the cost of spewing carbon dioxide, removing the primary ecological "subsidy" enjoyed by conventional fossil fuel plants, change could come quickly. With eco-pricing, wind power, for one, would beat coal dollar for dollar.[19] That would drive up the volume of sales in wind turbines, thereby drawing in the capital needed to commercialize next-generation technologies such as The Wind Turbine Company's. The same updraft would drive photovoltaics manufacturers towards the technologies whose chic look belies suitability for large-scale, low-cost production.

We'd probably still get some new coal plants, too. But at least they'd be clean.

A HISTORY OF ENERGY (CONTINUED)

1882 Thomas Edison opens the Pearl Street Power Station in New York
City, powered by an Armington & Sims high-speed steam engine.
This power plant, the world's first and the prototype for the central
power station, supplies power to about 2.6 square kilometres of
Lower Manhattan and provides service for eighty-five Wall Street
customers for roughly twenty-four cents U.S. per kilowatt-hour.

1883 The first public railway opens in Brighton, England. This railway will
continue running to the present day.

1884 The Englishman Charles Algernon Parsons builds the first steam
turbine to convert steam directly into electricity. The device, a
multi-stage reaction turbine, is used primarily to power ships. In
1896, the ship *Turbinia* will break the world speed record at sea
using Parsons's turbines, and will lead to the construction of
turbine-propelled warships for the British navy. Among Parsons's
other inventions is a non-skid device for automobile tires.

1888 The inventor and businessman Charles Francis Brush builds the
first automatically operated wind turbine for electricity generation
behind his Cleveland mansion. The enormous machine with 144
rotor blades made of cedar wood can only generate 12 kilowatts.
His company, Brush's Electric Company, will later be amalgamated
with Edison General Electric to form General Electric. In an effort to
make tricycle riding more comfortable for his small son, the Scottish
veterinarian John Dunlop invents the first air-filled or pneumatic
tire. His invention will soon catch on and turn bicycle riding into an
international phenomenon by the 1890s, and will soon be adopted
by the nascent car industry.

1891 Henry Ford, now an engineer for the Edison Illuminating Company,
designs a small engine that burns gasoline. Three years later he will
debut his first self-propelled vehicle, the "Quadricycle," with plant-
based ethanol powering the engine. In 1903, after a few abortive
attempts to raise investment, Ford will form the Ford Motor Company
with the vision to build a vehicle that is affordable to the working
family. Ten workers will be hired at a salary of US$1.50 per day
and Ford will begin developing the Model A.

▼ CONTINUED ON PAGE 210

08.

BOOM, BUST, AND EFFICIENCY

L. HUNTER LOVINS, WYATT KING

EDITORS' INTRODUCTION

When California passed a law in 1998 to open the state's electricity and energy market to competition — and to become the first state to privatize energy — it looked like a smart political strategy. Residential and business consumers were told that their energy costs would decrease.

By May 2000 the bold promises of deregulation were not playing out well. Instead of consumers getting cheap, reliable energy, the very opposite was happening. Caught in an early heat wave, California committed to what's called a Stage Two Emergency, signalling that backup power had dropped to less than 5 percent of demand. Some industrial customers were asked to reduce use to try to prevent a Stage Three, which calls for rolling blackouts to preserve the integrity of the system. The measures weren't enough. By June 15 the system had deteriorated further. A record-breaking heat spell caused the California Independent System Operator — basically the state's central control of energy — to implement "rotating localized outages" to preserve power for emergency operators such as hospitals. The blackouts caused panic, anger, and, as always, a lot of finger pointing. The governor blamed the power companies, who he claimed were colluding to take power out of the system through maintenance operations, in order to drive the price of energy skyward and increase profits. The companies denied it, blaming botched deregulation plans.

Throughout the rest of 2000 and early 2001, rolling blackouts and energy hikes became standard fare for Californians. Private energy utilities were forced to maintain price caps for the energy they were selling to their customers, while the price for the energy they were buying skyrocketed. A combination of irrational behaviour, higher demand for energy, and rapid deregulation was a disaster and led to bankruptcies. In a panic, Dick Cheney, the U.S. vice president, announced numerous new power plants, many of them coal-fired, would have to be built in order to supply the insatiable demand for energy. And then something unexpected happened: the California energy crisis that was supposed to last for decades was solved in a matter of months.

The answer everyone was looking for was simple: efficiency. As Hunter Lovins tells us in the next chapter, efficiency and human behaviour have always been a better answer to our energy problems then any new energy supply. In California the hike in energy prices led to major conservation efforts by the public. By June 2001 peak demand had fallen 12 percent from a year earlier, saving the equivalent of 4,800 megawatts, or the output of ten giant power plants. The vice president put his building plans on hold.

As founder of the Rocky Mountain Institute, and now as the president and founder of the Natural Capitalism Group, Hunter Lovins is one of the most sought-after experts on energy in the United States. As co-author of a number of extremely influential books, including *Natural Capitalism*, she has been at the vanguard of those looking at the true costs of our current energy systems. Not content merely to play the role of critic, she has been notable in her efforts to provide other, better models for how we might use energy in the future. In this chapter, along with her colleague Wyatt King, she makes her case for efficiency as the energy supply source of the future and why that may pose some problems for other power sources.

Promoters of clean energy often feel like Sisyphus, perpetually pushing a boulder up a mountainside. Just when solar, wind, and other clean technologies[1] seem poised to gain a foothold in the market, the ground collapses and the effort must begin again. This is not a new phenomenon. In the 1930s solar panels graced 30 percent of houses in Southern California and Florida. Several decades later few remained.[2]

Is the quest for clean energy doomed, or is an economy fueled by clean energy possible? What will it take to get over Sisyphus's hump?

It will require, first of all, an understanding of the role of energy efficiency, not merely a devotion to the various forms of renewable supply. The best way to understand energy efficiency is to take an end-use-least-cost approach and ask such questions as, What tasks need energy? and How can we perform those tasks in the cheapest and best ways? This approach will demonstrate that any form of clean energy supply likely will be the second-best option, and will be part of a durable answer only if integrated with the *best* choice: using less of the energy we already have.[3] Efficient use of energy should be considered before any supply technology, whether the supply is renewable, nuclear, or fossil. Unless the relationship between energy efficiency and supply is understood, any effort to convert to secure, affordable, abundant supplies of energy is likely to fail. Efficiency will then play the role of spoiler: dampening any supply strategy. It will defeat renewable energy, as surely as it has laid waste to most traditional supply technologies.

HOW NOT TO MAKE ENERGY POLICY

Official government energy strategies have tended not to work. This is, in large part, because they have tried to dictate to the market technologies that cannot survive without government support. In consequence, the policies have been more honoured in the breach than in the observance. In the 1950s the Paley Commission (the first U.S. government commission to study energy and recommend policies to ensure secure, affordable supplies of future energy) called for a massive conversion to renewable energy, but its sensible and timely recommendations were completely ignored. The world seemed awash in oil at the time.

The 1973 oil embargo changed that perception. It also gave President Nixon the dubious honour of having to put forth an official energy policy. His proposed Project Independence would have had Americans spend three-quarters of all discretionary investment money in the economy on new forms of energy supply, disproportionately coal and nuclear. The market, however, took a dim view of this plan, and few of these investments ever happened. Nixon's response was to deal with the inflation caused by the run up in energy costs by capping energy prices. This move, however, denied institutional and individual decision makers the market mechanism of a price signal.

In due course, the oil markets settled down and few people beside the usual suspects of electric utility companies, oil executives, and a few beady-eyed policy wonks spent much time worrying about how the country would meet its needs for energy. A growing number of citizens, though, started to concern themselves with the consequences of the Western world's energy use. The environmental movement, in particular, launched protests against the conventional supply technologies and called for a conversion to a solar economy.

Some people thought that all that was needed was a change of administration, but when President Carter had to respond to his own energy crisis in 1979, not surprisingly his solution also featured governmental directives and a disproportionate emphasis on conventional supply. To overcome the market's continued reluctance to pay for such a prescription, Carter proposed the Energy Security Corporation and the Energy Mobilization Board. These agencies could override market mechanisms and supplant democratic institutions if either worked to impede supply expansions. In an effort to pacify environmentalists, Carter put solar panels on the White House, and his Department of Energy took a liking to every sort of centralized solar technology imaginable, from solar space satellites to solar power towers in the desert to wind machines with blades the length of a jumbo jet wing.

Once again, though, the effort to "solve" the energy problem with central mandates failed. Even the declaration that the country's energy shortages were the moral equivalent of war did not help Carter's energy program, which was so capital intensive that it failed a test of market rationality.

On the positive side, Carter was the first president to recognize the advantages of energy efficiency. His administration implemented such measures as CAFE (Corporate Average Fuel Efficiency) standards for vehicles. These were regulatory programs, but they did focus on eliciting the cheapest solutions to the energy problem. They also dramatically increased American security by enabling the country to buy less oil more quickly and on a larger scale than the Organization of Petroleum Exporting

Countries (OPEC) could adjust to. New U.S.-built cars increased their efficiency by 7 miles per gallon (mpg; 33.6 litres per 100 kilometres [L/100 km]) in six years. Europe achieved similar oil consumption savings, but did so primarily through higher fuel taxes rather than through better efficiency standards.

Together, these changes tipped the world oil market in buyers' favour. Between 1977 and 1985, U.S. oil imports fell 42 percent, depriving OPEC of one-eighth of its market. The entire world oil market shrank by one-tenth; the United States alone accounted for one-fourth of that reduction. OPEC's share fell from 52 percent to 30 percent, forcing it to cut its output by 48 percent, which drove down world oil prices. On average, new cars each drove 1 percent fewer miles, but used 20 percent fewer gallons. Only 4 percent of those savings came from making the cars smaller.[4]

Carter's plans also put in place two measures that enabled the market to work better: they lifted the market caps on price

DESPITE ITS IMPERFECTIONS, THE MARKET, GIVEN HALF A CHANCE TO WORK, TURNED OUT TO BE SMARTER THAN THE SUPPOSED ENERGY EXPERTS.

and they initiated a variety of programs that provided information on what sorts of energy technologies were available and what citizens could do to use energy more wisely.

Carter's efficiency initiatives worked far better than his supply plans, and their beneficial effects lingered for half a decade after his term. Between 1979 and 1986, Americans cut total energy use by 5 percent — a drop in energy intensity (the amount of energy needed to produce a unit of GDP) that was five times bigger than the expanded coal and nuclear output subsequently promoted by President Reagan's policy.

Upon entering office in 1981, Reagan sought to stimulate fossil fuel and nuclear energy production, not realizing that the efforts of the previous administration had allowed the United States to cut energy intensity at the record pace of 3.5 percent per year. Five years later, energy efficiency — disdained as an intrusive sacrifice and a distraction from America's supply prowess — had pre-empted the markets that were supposed to pay for costly supply expansions. In the mid-1980s, many of the producers Reagan meant to help were ruined, as efficiency's speed and availability made energy prices crash. Despite Reagan's concerted campaign to undo the previous administration's efficiency and

ENERGY EFFICIENCY CAME ONLINE FAR FASTER THAN ANYONE HAD PREDICTED, AND FAR FASTER THAN ANY EXPANSION OF SUPPLY.

information programs,[5] by the middle of the decade the market had had time to work. Entrepreneurs were introducing myriad technologies that were producing huge efficiency benefits. Even advocates of renewable supply were caught off guard, as well as being hampered by the inept way that government programs sought to subsidize renewables. It turned out that efficiency was simply much cheaper than any form of supply. Despite its imperfections, the market, given half a chance to work, turned out to be smarter than the supposed energy experts.

Energy efficiency came online far faster than anyone had predicted, and far faster than any expansion of supply. From 1983 to 1985, the nation's third-largest investor-owned utility was cutting its decade-ahead forecast of peak demand by about

8.5 percent *each year*, at roughly 1 percent of the cost of new supply. The nation's largest investor-owned utility signed up 25 percent of new commercial construction projects for design improvements in just three months. As a result, it raised its target for the following year — and hit it in the first nine days. Well-designed efficiency programs have captured up to 99 percent of target markets. A huge literature confirms the size of the savings and shows that the costs of achieving them can be accurately predicted and measured.[6]

This history repeated itself in 2001 as President George W. Bush, with ardour similar to Reagan's, sought to stimulate energy supplies, even though in 1996 the United States had quietly resumed saving energy by 3.2 percent a year. Bush called for opening the Arctic National Wildlife Refuge to oil exploration, and proposed massive fossil and nuclear subsidies. Meanwhile, subsidies and other policies that promoted vehicle inefficiency led to a situation where the average fuel efficiency of U.S. cars and trucks hit a twenty-two-year low in 2002: 20.4 mpg (11.5 L/100 km).[7] In June 2003, environmentalists pointed out that the average fuel efficiency of Ford cars and trucks was worse than when the company started one hundred years ago with the Model T.[8] This represents a tremendous lost opportunity. The U.S. National Academy of Sciences reported in 2001 that although light-vehicle improvements have already cut gasoline consumption by 14 percent — equivalent to the amount of oil imported from the Persian Gulf — further efficiency gains, which would be cost-effective to the driver to make, can roughly double U.S. fleet efficiency without compromising safety or performance.[9] Typical potential fuel savings range from about one-fifth for small cars to one-third for midsize SUVs and nearly one-half for big pickup trucks. Achieving such savings would be good for more than driver's pocketbooks: such vehicles are responsible for 20 percent of U.S. carbon dioxide emissions.

Such savings projections are quite conservative and assume that smarter car companies will not introduce novel designs that will disrupt the industry and put inattentive car companies at risk. While American car companies resist making their products more fuel efficient, the Japanese and Europeans are again designing the future. The Toyota Prius hybrid-electric five-seater gets 48 mpg (4.9 L/100 km); Honda's Insight gets 64 mpg (3.7 L/100 km). An entire American car fleet that efficient would save thirty-two times the amount of oil that proponents of drilling in the Arctic hope to find there.[10] DaimlerChrysler and General Motors are now testing family sedans at 72–80 mpg (3.3 L/100 km–2.9 L/100 km). Volkswagen already sells Europeans a 78-mpg (3.0-L/100-km) four-seat non-hybrid subcompact. Almost every automaker at the 2003

ONCE THE WORLD LEADER IN ENERGY EFFICIENCY, WITH FINANCIALLY HEALTHY UTILITIES AND SENSIBLE RESOURCE POLICIES, CALIFORNIA NEARLY PLUNGED THE WHOLE OF THE UNITED STATES INTO THE NEXT ENERGY CRISIS.

Tokyo Auto Show displayed good hybrid-electric prototypes, some getting more than 100 mpg (2.35 L/100 km). Volkswagen has just premiered a supersafe but ultralight diesel car that gets 285 mpg (0.83 L/100 km).[11]

Markets are, of course, imperfect. As prices fall, people easily fall back into apathy. Advertising campaigns (and tax subsidies) that encourage Americans to buy a 10-mpg (23.5-L/100-km) Hummer H2 so that they can paste an American flag on it and feel like they are patriotically supporting the troops in the Middle East, ensures that those young men and women will continue to be in harm's way, driving 0.5-mpg (470-L/100-km) tanks and 17-feet-per-gallon (73,000-L/100-km) aircraft carriers.

Such behaviour also ensures that we will all get to enjoy yet another energy crisis.

In 2000, California created such a crisis with its ham-handed program to "deregulate" electricity markets in such a way as to allow the incumbent "big dogs to eat first."[12] Once the world leader in energy efficiency, with financially healthy utilities and sensible resource policies, California nearly plunged the whole of the United States into the next energy crisis. Panicked by so-called power shortages and exhorted by Vice President Dick Cheney's call to build at least one power plant a week,[13] developers planned to add electricity-generating capacity equivalent to 83 percent of the state's current total demand, 96 percent of the western region's, and at least one-third of the nation's. But meanwhile, California citizens, companies, and communities woke up and implemented exactly the same solution that had worked before: reliance on efficiency, which enabled them to save their way out of the hole. Californians cut peak electricity demand per dollar of gross domestic product (GDP), adjusted for weather, by 14 percent in six months, and a third of customers cut their usage by over 20 percent. In just the first six months of 2001, customers wiped out California's previous five to ten years of demand growth, taking away proposed new power plants' markets before their plans could even be finished. This abruptly ended the crisis that the White House claimed would require 1,300 to 1,900 more power plants nationwide. By August 2001, a *Barron's* cover story noted a coming glut of electricity. Scores of plants have been cancelled for lack of demand,[14] and their irrationally exuberant builders are reeling as Wall Street, stung by Enron's collapse, downgrades their bonds.

Efficiency keeps quietly making gains. A 2002 report by the American Council for an Energy-Efficient Economy (ACEEE), "State Scorecard on Utility and Public Benefits Energy Efficiency Programs: An Update," states that there has been a halt to the decade-long trend to eliminate or reduce efficiency programs in the United States. Spending on utility and related state energy programs has rebounded modestly from the late 1990s. Annual spending on energy efficiency programs reached a high point of about US$1.6 billion in 1993 and dropped dramatically to about US$900 million in 1997. This resulted from the spread of utility deregulation in the mid-1990s. Revived interest in energy efficiency by the states has begun to reverse that trend, with total spending by states and utilities on energy efficiency programs back to about US$1.1 billion in 2000. "Our analysis clearly illustrates that there remains a vast resource of energy efficiency opportunities in the United States that is being largely ignored and untapped," stated Dan York, ACEEE's Utilities Research Associate and co-author of the report. "Most states still offer no significant support for efficiency programs, and federal energy legislation has so far ignored the need for a national matching funding mechanism for state efficiency programs. This leaves the main burden of support for efficiency programs to a few states."[15] Since that report, however, at least three states have voted to implement efficiency programs as a way to cut emissions of carbon dioxide.

EFFICIENCY IS THE REAL ENERGY SUPPLY TECHNOLOGY

In each energy crisis that has beset the world, efforts to promote one or more forms of energy supply have ignored the role of efficiency in creating a successful energy strategy. As if to spite all the brainpower and paper thrown at the energy problem (some wags suggest that the solution to the energy problem is to burn energy studies), the manifestly imperfect market has quietly bought more efficiency than it has new forms of supply. Markets are motivated by price, information, and consumer values. After 1979 there was a real perception of crisis. Prices spiked. People sought information.

And when the government, utilities, and various non-profits supplied it, the market mechanisms worked rapidly to "solve" the energy crisis. Efficiency swamped supply and prices crashed.

    This persistent oscillation, driven by what could be called "the overhang of profitable efficiency," has repeated itself at least

EVERY TIME PRICE HIKES, APPARENT SHORTAGES, OR POLITICAL INSTABILITY CREATE THE PERCEPTION THAT THE TIME IS RIGHT TO CONVERT TO A DIFFERENT FORM OF ENERGY SUPPLY, EFFICIENCY, THE CHEAPER AND QUICKER ALTERNATIVE, ELIMINATES THE PERCEPTION OF A NEED FOR CHANGE AND ALLOWS THE ENERGY STATUS QUO TO RESUME.

four times since the 1973 Arab oil embargo, and will do so again. Every time price hikes, apparent shortages, or political instability create the perception that the time is right to convert to a different form of energy supply, efficiency, the cheaper and quicker alternative, eliminates the perception of a need for change and allows the energy status quo to resume. This fuel bazaar continues to result in bankrupt supply companies, a climate that grows less stable by the year, energy vulnerability, and war in the Middle East.

    Few people other than those whose careers depend on energy analysis really want to focus on how they get their energy. If prices remain relatively low (and the world price of oil is still below that of bottled water) few people will overcome the "hassle factor" to make a change in their energy system. Every time the price for the prevailing energy source (usually oil) gets high enough to make people interested in overcoming that hurdle, the myriad ways of using energy more efficiently become more attractive. Given that it is usually faster to implement efficiency than to bring in new forms of energy supply, efficiency outpaces all supply options. This reduces demand for energy, drives prices down, and dissolves any sense of vulnerability. Proponents of supply are back at square one, the falling price of oil having diminished the relative attractiveness of their pet technologies.

Avoiding this boom-and-bust cycle requires understanding its three root causes. First, improved efficiency costs far less than new energy supplies, so most people, given the choice, opt for efficiency. Second, when policies promote *both* efficiency and supply, customers typically use only one, usually the cheaper one. Third, efficiency is far *faster* than new supply. Ordinary people are able to implement efficiency long before big, slow, centralized plants can be built, let alone paid for.

Since Western economies ceased to think that oil was infinite and reliably available, more efficient use has been the biggest "source" of new energy — not oil, gas, coal, or nuclear power. After the 1979 oil shock, efficient use of energy enabled

IN NEARLY EVERY CASE, ENERGY EFFICIENCY COSTS LESS, USUALLY FAR LESS, THAN THE FUEL OR ELECTRICITY THAT IT SAVES.

Americans to cut oil consumption by 15 percent in six years while the economy grew 16 percent. There are many ways to measure progress in saving energy but even by the broadest and crudest measure — lower primary energy consumption per dollar of real GDP — progress has been dramatic. If the energy use of 1975 is taken as a base measure, by 2000, reduced "energy intensity" was providing 40 percent of all U.S. energy services. It was 73 percent greater than total U.S. oil consumption, five times greater than domestic oil production, three times greater than all oil imports, and thirteen times greater than Persian Gulf oil imports. The lower intensity was mostly achieved by more productive use of energy (such as better-insulated houses, better-designed lights and electric motors, and cars that were safer, cleaner, more powerful, and got more miles per gallon). The savings were only partly caused by shifts in the economic mix, and only slightly by behavioural change. Since 1996, saved energy has been the nation's fastest-growing major "source."[16]

In nearly every case, energy efficiency costs less, usually far less, than the fuel or electricity that it saves. It is cost effective to save at least half the energy now used in developed countries at prices averaging around 2¢ per kilowatt-hour (kWh).[17] Almost no form of new supply, and few historic ones, can compete with this.

The 40 percent drop in U.S. energy intensity since 1975 has barely dented efficiency's potential. The United States has cut annual energy bills by about US$200 billion since 1973, yet is still wasting at least US$300 billion a year. That number keeps rising as smarter technologies deliver more and better service from less energy. And the side benefits can be even more valuable; for example, studies show 6 to 16 percent higher labour productivity in energy-efficient buildings.[18]

The huge potential of efficiency can contribute to a transition to a clean energy economy, but this is unlikely to happen unless national energy policies integrate strategies to implement efficiency as a conscious part of implementing other clean technologies.

RENEWABLE POWER TECHNOLOGIES

We are witnessing a renewed boom in renewable energy, as advocates of clean energy call ever more loudly for a transition to clean energy. The following table illustrates this point:

TRENDS IN ENERGY USE, BY SOURCE, 1995–2001

Energy Source	Annual Rate of Growth (percent)
Wind power	+32.0
Solar photovoltaics	+21.0
Geothermal power*	+4.0
Hydroelectric power	+0.7
Oil	+1.4
Natural gas	+2.6
Nuclear power	+0.3
Coal	–0.3

*Data available through 1999.[19]

Europe is perhaps the best known of the regions pursuing renewable supply. Europeans have a much clearer understanding than North Americans that climate change is real and that effective policy measures are needed to counter it. In 2001, the European Union decreed that 22 percent of electricity, and 12 percent of all energy, should come from renewable sources such as wind within ten years. This target is part of the way the E.U. intends to meet its obligations under the Kyoto Protocol.[20]

According to a study released in 2002 by the European Wind Energy Association (EWEA), Europe's wind energy industry grew by 40 percent over the last year. In the twenty-one countries included in the study, installed wind capacity rose from 14,652 megawatts to 20,447 megawatts (MW) between October 2001 and October 2002. According to the same study, capacity on the continent could rise to 100,000 MW by the end of the decade. The European wind power industry estimates that, given the right legal and financial support, wind projects could provide energy for 50 million people in Europe in less than ten years' time.[21]

Following on the German government's decision in the late 1990s to phase out nuclear power completely, Germany has begun to pursue the most dramatic expansion of renewable energy in the world. In recent years Germany has accounted for roughly half of all wind turbines built worldwide. Bundesverband WindEnergie, the German wind energy association, recently announced that 2002 was another record-breaking year for installation of wind energy systems in that country. A total of 3,247 MW of generating capacity were installed last year, bringing German wind supplies to more than 12,000 MW, produced by nearly 14,000 wind turbines. Four and a half percent of German electricity in now generated from wind, surpassing the contribution from hydroelectric power. The wind sector in Germany now employs 45,000 people, one-fifth of whom were hired last year.[22] This rapid growth in wind power is central to reaching Germany's goal of reducing carbon emissions 40 percent by 2020.

Authorities in Germany are now considering plans to build up to 5,000 turbines off the country's north coast. Giant turbines, double the size of conventional ones, are being developed for this use. Some of the turbines would be located in the open sea up to 45 kilometres from land — a feat never before attempted. Since wind is stronger at sea, the energy potential is highly attractive. Already the world's leading country in the development of onshore wind energy, Germany has plans to add 25,000 MW to offshore capacity by 2030, up from the current level of zero.[23]

Close behind Germany in installed wind power is Spain, which currently ranks number two in Europe with 4,079 MW installed capacity.[24]

The press in the Unites States is beginning to take notice of such developments. According to an article published in *USA Today* on February 7, 2002,

Throughout Europe, wind power has turned into a serious source of energy, leaving the USA — the country that pioneered it as a modern technology — in the dust. Amid growing concern about climate change and other environmental problems blamed on the burning of fossil fuels, European governments are encouraging utility companies to harness the wind, especially at sea where it blows hardest.

In 2001, EU countries produced more than four times as much energy through wind as the USA, and experts predict that within 10 years at least 10 percent of Europe's electricity will be supplied by giant wind turbines hooked up to main power grids. Even the technology used to produce power from wind, originally a US development, has moved to Europe. GE is the only company that still makes wind turbines in the US — 90 percent are now produced in Europe. According to Randall Swisher, executive director of the American Wind Energy Association, "We have frittered away our dominant role in this technology . . . We had the strategic advantage, and we lost it."

While the United States is not leading the wind revolution, it is also not ignoring it. U.S. wind-generating capacity expanded by nearly 10 percent in 2002, to a total of 4,685 MW. However, development depends largely on the existence of a federal wind tax credit, which must be renewed every two years. Growth in 2002 was slower than in previous years due to the fact that the tax credit had expired at the end of 2001 and an extension was not signed until mid-March. During those first months of 2002, many wind development projects were placed on hold.[25]

Environmental researcher and author Lester Brown of the Earth Policy Institute writes,

Over the last decade wind has been the world's fastest-growing energy source. Rising from 4,800 megawatts of generating capacity in 1995 to 31,100 megawatts in 2002, it increased a staggering six-fold worldwide. Wind is popular because it is abundant, cheap, inexhaustible, widely distributed, climate-benign, and clean — attributes that no other energy source can match. The cost of wind-generated electricity has dropped from 38¢ a kilowatt-hour in the early 1980s to roughly 4¢ a kilowatt-hour today on prime wind sites. Some recently signed U.S. and U.K. long-term supply contracts are providing electricity at 3¢ a kilowatt-hour. Wind Force 12 projected that the average cost per kilowatt-hour of wind-generated electricity will drop to 2.6¢ by 2010 and to 2.1¢ by 2020. U.S. energy consultant Harry Braun says that if wind turbines are mass-produced on assembly lines like automobiles, the cost of wind-generated electricity could drop to 1–2¢ per kilowatt-hour. In contrast with oil, there is no OPEC to set prices for wind. And in contrast to natural gas prices, which are highly volatile and can double in a matter of months, wind prices are declining. Another great appeal of wind is its wide distribution. In the United States, for example, some 28 states now have utility-scale wind farms feeding electricity into the local grid. While a small handful of countries controls the world's oil, nearly all countries can tap wind energy.[26]

Worldwide, wind grew by a robust 36 percent in 2001 alone.

Renewables are making headway in developing countries as well. The People's Republic of China is undertaking a rapid switch from coal to gas and investing in efficiency measures and renewable energy sources, pushed by the need to boost

economic development and reverse the public-health emergency caused by air pollution. In 1996, China mined 1.4 giga-
tonnes (GT) of coal. Most experts thought that would double
early in the new century. But in 2002 China's coal mining was
back to its 1986 level — 0.9 GT — and heading for 0.7 GT. A
modern natural gas infrastructure is being built with wartime
urgency in five key cities. Modern Danish wind turbines are
being installed in Mongolia. China, which cut its energy intensity of economic growth in half in the 1980s, has nearly done
so again, and can do more. Its transition is driven not only by
the fact that coal is unacceptably dirty, but also by the realization that if coal remains the primary fuel for the country's
development, there would be no rail capacity to carry anything
but coal. In addition, the Chinese are taking note of hybrid-
electric cars, fuel cells, and hydrogen, and they are becoming
very interested in the entire concept of sustainability. The first
run of the book *Natural Capitalism* sold out in two days, and
they have recently created a Department of Sustainability at
Peking University. The Chinese have also been active participants with Royal Dutch Shell in developing energy planning
scenarios.[27]

Similarly, India is one of the world leaders in wind energy,
adding 240 MW of wind capacity in 2001. As of 2002 it had 1,627
MW of installed wind turbines.

According to Lester Brown,

Projecting future growth in such a dynamic industry is
complicated, but once a country has developed 100
megawatts of wind-generating capacity, it tends to move
quickly to develop its wind resources. The United States
crossed this threshold in 1983. In Denmark, this occurred
in 1987. In Germany, it was 1991, followed by India in
1994 and Spain in 1995.

By the end of 1999, Canada, China, Italy, the Netherlands, Sweden, and the United Kingdom had crossed this threshold. During 2000, Greece, Ireland, and Portugal joined the list. And in 2001, it was France and Japan. As of early 2002, some 16 countries, containing half the world's people, have entered the fast-growth phase.[28]

Projections of worldwide use of renewables follow what is happening at the country level. In 1998 the Royal Dutch Shell external relations newsletter, "Shell Venster," stated that "in 2050 a ratio of 50/50 for fossil/renewables is a probable scenario, so we have to enter this market now!" Shell's "Dynamics as Usual" scenario finds it plausible that renewables will supply 20 percent of world energy by 2020, and a third by 2050. Their more aggressive scenario, "Spirit of the Coming Age," finds a transition to a hydrogen economy plausible by 2050, driven in part by a Chinese conversion to hydrogen.[29] In 1995 London's Delphi Group began advising its institutional investment clients that alternative energy industries offer "greater growth prospects than the carbon fuel industry."[30]

The U.S.-based Solar Energy Industries Association said that solar research has cut prices to a point where the world could expect to see photovoltaic panels competing with natural gas–fired generation within the next five to eight years. Statistics from the Global Environment Facility show that the market for photovoltaic solar energy is growing by 15 percent a year. The United Nations' World Energy Assessment said solar thermal power plants covering just 1 percent of the world's deserts could meet the entire planet's current demands for energy.[31]

Japan leads the world in installed solar generating capacity with approximately 400 MW. Installed solar power in Germany stood at 200 MW in 2001, while the United States ranks third with approximately 179 MW in installed solar capacity.[32]

HOW MANY ECONOMISTS DOES IT TAKE TO CHANGE A LIGHT BULB?[33]

Efficiency is well-established and has favourable economics. Renewables are on a rapid-growth path. Shouldn't the market simply sort all this out?

Perhaps — if we had a true market, free of distorting subsidies. Unfortunately, nowhere does such a market exist in energy or any other commodity. Energy choices around the world are beset by subsidies and market distortions of all sorts, which have, in large measure, dictated the energy mix that we have today. Worldwide, it is estimated that subsidies to the energy sector, overwhelmingly to fossil fuels, top US$240 billion each year.[34] Any strategy that seeks to foster a transition to clean energy has to reckon with these distortions. Unfortunately, few government officials even acknowledge they exist.[35] This is especially true in Europe, where only recently have any competent estimates been made of subsidies to the energy sector.

In the United States, historic subsidies to nuclear power, for example, have exceeded the money spent on the Vietnam War and the space program combined. This to deliver less energy than the burning of wood. According to one estimate, U.S. government subsidies to the energy sector as a whole are at least US$30 billion per year, a disproportionate amount of which goes to support the nuclear and fossil fuel industries.[36] Because of this, "The American economy is, after Canada's, the most energy-dependent in the advanced industrialized world, requiring the equivalent of a quarter ton of oil to produce [US]$1,000 of gross domestic product. Americans require twice as much energy as Germany — and three times as much as Japan — to produce the same amount of GDP."[37]

One reason renewables have had such a hard time gaining a foothold in the United States is that they compete not only with subsidized conventional energy, but also with efficiency. Recently, the U.S. Department of Energy (DOE) reported that the use of renewable energy fell in 2001 to its lowest level in twelve years. Much of that was due to low hydroelectric output from reduced snow pack in the western states, but the DOE noted that solar generating equipment was also being retired faster than it was being replaced.[38]

This is all clearly daft. It is also a recipe for uncompetitiveness. But in light of the 2001 Cheney energy proposals (which remain largely a gleam in the administration's eye), it is perhaps unreasonable to look to the Bush administration to provide a level playing field on which efficiency and all forms of supply might compete fairly. The administration even took money from the Energy Department's solar and renewable energy and energy conservation budgets to pay for the cost of printing its national energy plan, which called for reducing such programs and increasing subsidies to fossil and nuclear technologies. Reuters reported that "documents released under court order by the Energy Department this week revealed that [US]$135,615 was spent from the DOE's solar, renewables and energy conservation budget to produce 10,000 copies of the White House energy plan released in May 2001."[39]

Despite public opinion polls showing support for renewable energy, there is also growing resistance to particular applications. The citizens of Cape Cod are fighting a proposed wind farm for Nantucket Sound, the first offshore facility in the United States. The *New York Times* reported, "But like residents of dozens of communities where other wind-farm projects have been proposed, many Cape Codders have put aside their larger environmental sensitivities and are demanding that their home be exempt from such projects. As (Walter) Cronkite puts it, 'Our national treasures should be off limits to industrialization.'"[40] The proposed wind farm is on hold.

There is also some question about how the electricity from

the wind farms would get to market. While farmers and ranchers throughout the Heartland typically welcome wind farms as great neighbours to their cows and corn fields, the communities through whom the transmission lines would have to pass to carry the wind energy to distant power-hungry cities are considerably less enthusiastic. And it is not entirely clear where the money for the transmission lines will come from. Such capital costs will raise the cost of the *delivered* power. Once again, efficiency may come to look increasingly attractive.

THE BEGINNINGS OF AN INTEGRATED POLICY

While American energy policy is drafted by promoters of technologies beloved in the oil patch, the Europeans are beginning to realize that an integrated strategy of efficiency and renewables might just enable them not only to get beyond the historic boom-and-bust oscillations, but also give them a competitive edge.

In 1999, then British environment minister Michael Meacher said, "I cannot over-emphasise that improved energy efficiency, and growth in renewable energy, are not alternatives — we need to pursue both issues vigorously, and we are doing so." And the deputy prime minister, John Prescott, announced the Climate Change Programme in March of 2000 by saying, "We need a radical shake up in the way we use energy and we need to generate energy in new sustainable ways."[41]

An integrated policy is vital not only because ignoring efficiency will endanger a renewable (or conventional) supply strategy, but also because focusing first on efficiency makes any supply strategy much more attractive. In the absence of energy efficiency, supply strategies become prohibitively expensive. With efficiency, renewables can provide far less supply, and do so more cost-effectively than conventional power.[42] A dollar can only be spent once. If it buys efficiency, the best buy, more of our budget is left to buy the increasingly attractive renewable supply options. If that dollar is spent on centralized, capital-intensive conventional supply, it cannot then be spent to save the energy that will make much of that supply unnecessary — until the higher prices that will be necessary to pay back the invest-ments in conventional supply elicit defensive investments in effi-ciency. But it is exactly this sort of cycle that has ensured energy insecurity. The only answer, as the Europeans are starting to realize, is to invest first in efficiency, then in renewables, and to do so as part of a conscious, integrated plan.

A 2000 European Commission Green Paper, "Towards a European Strategy for Energy Supply Security," highlighted a central role for energy efficiency in increasing the security of supply and reducing greenhouse gas emissions. It stated that improving the efficiency with which energy is consumed by end users is central to European energy policy, since improved efficiency meets all three goals of energy policy, namely secu-rity of supply, competitiveness, and protection of the environ-ment. This is further developed in the European Climate Change Programme, which highlights the large potential for cost savings from improving the energy efficiency of end-use equipment.[43]

Recent pronouncements go even further. The Energy-Intelligent Europe Initiative is a cross-party, cross-nation movement within the European Parliament that calls for making Europe's economy the most energy intelligent in the world. By February 15, 2002, forty-one parliamentarians from all fifteen members states had signed the call to promote energy efficiency in Europe as the number one energy "source." Linking energy intelligence to the knowledge-based economy "will help Europe to become the most compet-itive economy worldwide while achieving its ultimate goal, a sustainable development." The initiative concludes that energy efficiency is not perceived as an important policy tool at the moment, but points out that a more energy-intelligent economy is what will enable Europe to remain competitive and promote a high quality of life.[44]

But critics are skeptical of such initiatives, claiming that funds for the promotion of energy efficiency remain inadequate: "The entire budget will amount to just over 1 million Euros per member state per year . . . a minor percentage increase upon budgets originally set well over a decade ago, when not even lip-service was being paid to the need to prioritize sustainable energy." They argue that "it is appropriate for the Union to concentrate on guiding and steering demand, unlike the United States, which seeks to meet demand by constantly boosting supply."[45]

THE ECONOMIC BENEFITS OF AN INTEGRATED STRATEGY OF EFFICIENCY AND RENEWABLES

What would an intelligent combination of efficiency and renewable supply look like? It turns out that it has been done, and the combination offers a winning way to strengthen local economies and create new jobs. The example also shows that even if national governments continue to be unable to grasp this concept, there still remains hope at the local level.

Sacramento is California's capital city, with a population of 400,000 (in a metro area of 1.8 million). The Sacramento Municipal Utility District (SMUD) demonstrated how investments in efficiency *and* locally generated power can enhance the bottom line of the utility and improve the health of the regional economy.

SMUD is the sixth-largest municipal utility in the United States, serving 1.2 million customers.[46] In 1989 its customers/owners voted to close the Rancho Seco nuclear facility. According to Jan Schori, who became general manager of the utility in 1993 and who remains at the helm today, "When we closed the Rancho Seco nuclear plant we lost 913 megawatts on a 2,100-megawatt system. It became an opportunity for us to start over."[47]

By 1995, SMUD was spending 8 percent of its gross revenue on energy efficiency and was being described "as a symbol of what's possible . . . the national poster child of green utilities."[48] This investment reduced the peak load by 12 percent and enabled SMUD to hold rates constant for ten years. Had Rancho Seco operated, rates would have increased 80 percent.

SMUD also installed

- the nation's largest photovoltaic power plant, providing 2 MW of solar power to 500 homes, located next to the closed Rancho Seco nuclear facility;
- one of the largest utility-owned commercial wind turbine projects in the United States, producing 5 MW;
- the largest solar home project in the nation — supporting one hundred customers a year with the installation of 4-kilowatt photovoltaic panel systems on their rooftops;
- one of only two photovoltaic recharging stations for electric vehicles in the United States;
- two geothermal projects with a total generating capacity of 134 MW.[49]

SMUD partnered with the Sacramento Tree Foundation to plant 300,000 shade trees from 1990–2000, and it continues to offer customers free trees (with advice, fertilizer, and free delivery) for planting on the east, west, or south side of buildings. Full-grown trees can reduce indoor cooling requirements by up to 40 percent.[50] The district has also helped customers to purchase over 42,000 superefficient refrigerators.[51]

Continuing its focus on cost-effectively reducing energy

demand, SMUD instituted a Cool Roof program. Building owners, through contractors, can earn a SMUD rebate of 20¢ per square foot for installing Energy Star sun-reflecting coating on flat roofs. The highly reflective coating helps block heat from the sun from being absorbed through a flat roof and into a building. This means less energy is consumed by air-conditioning systems.[52]

An economist calculated the present value of SMUD's 1997–2001 energy efficiency programs for the Sacramento region's economy to be US$130 million over the life of the efficiency investments. Most businesses are expected to save 10–19 percent on their energy bills, which translates into more jobs and profits, and increased wages and competitiveness. The impacts include the creation of over 150 additional job-years for a dozen years.[53] One company, which had anticipated higher rates that would force it to close, was able to stay in business, saving 2,000 jobs. Sacramento's competitive rates attracted such new factories as Apple, Intel, and a solar equipment manufacturer. The program lowered the utility's debt, upgraded its credit, and made it the most competitive utility in California.

By 2000, under SMUD's photovoltaic installation program, more than 450 residential and 30 commercial photovoltaic systems had been installed. These systems are grid-connected and feature net metering, which, by earning revenue from providing power to the electrical grid, allows SMUD to recuperate more than half of the cost of the systems.[54] The current program offers the systems to homeowners at a lower cost than private marketers do. The SMUD Web site says the systems provide "virtually free energy after an 8–15 year payback period." They are expected to have a thirty-year lifespan. Though they increase home values, no additional property taxes are levied on the value of the systems. As of September 2001, the California energy shortage has caused a tremendous surge in interest in rooftop solar systems, leaving SMUD with a large backlog of orders.[55]

About half of SMUD's current power supply comes from its own hydro, wind, and photovoltaic power plants (at 8 MW, SMUD uses more solar power than any other utility in the United States) and from four highly efficient natural gas cogeneration plants built in the 1980s and 1990s. The other half of its power supply is purchased through long-term contracts, and SMUD searches for the best market prices. Despite Sacramento's continued growth, SMUD helped shave annual peak power requirements by nearly 3 percent from 1999 to 2000.[56]

In 1999 the U.S. Environmental Protection Agency (EPA) office in Richmond, California (a Bay Area suburb), became the first federal facility entirely powered by green power through a contract with SMUD. During the first year of the contract between the EPA and SMUD, 60 percent of the building's power was sourced from geothermal energy, with the remaining 40 percent coming from a landfill's gas generation. In the future, all of the building's energy will come from the landfill.[57] Other such projects were implemented in 2000 and 2001.

In October 2001, SMUD's board of directors voted for a ten-year strategic plan developed by General Manager Schori. The proposed plan calls for

- saving enough electricity through energy efficiency to power more than 40,000 homes;
- maintaining competitive rates that are now 30 percent lower than those of the neighbouring utility, PG&E;
- adding new wind power to meet the needs of 12,000 homes and new solar power to serve up to 8,000 homes;
- building a new 500 MW combined-cycle gas-powered plant adjacent to the closed Rancho Seco plant. The new plant will meet a large portion of Sacramento's round-the-clock electricity demand and bolster SMUD's system reliability.

The plan diversifies SMUD's fuel mix, reducing the financial risks of relying on one fuel or generating source. "As we've seen in the past 18 months, no one can predict uncertainties such as prolonged dry water years and major shifts in market conditions," Schori said. "This is a progressive yet prudent plan for meeting Sacramento's long-term energy needs with one of the cleanest, most reliable and affordable energy mixes in the state."[58]

CONCLUSION

Albert Camus argued that Sisyphus was free because, though condemned by the gods forever to roll his rock, he could use his trip back down the hill for personal reflection. But for promoters of renewable energy supplies, this freedom comes at a terrible cost.

In North America, we are not learning from our steps backward. We do not bother to implement a smart energy policy

ENERGY EFFICIENCY SHOULD BE THE CORNERSTONE OF ANY ENERGY POLICY THAT HOPES TO SURVIVE THE RIGOURS OF THE MARKET.

or a timetable for moving away from conventional fuels. Energy efficiency should be the cornerstone of any energy policy that hopes to survive the rigours of the market. It is the cheapest way to meet the demand for the benefits that energy can deliver: hot showers and cold beer, the movement of goods and people, and the development of emerging economies. Coupled with renewable energy technologies, it can meet the needs of the world for energy services while supporting local community economic development. It is cheaper than any form of supply, and in a genuinely competitive market, it will render most proposals for new supply unattractive.

Unfortunately, efficiency remains largely ignored as an energy source in North America. At least until the next energy crisis.

A HISTORY OF ENERGY (CONTINUED)

1892 The French engineer Rudolf Diesel applies for a patent for his self-named pressure-ignited heat engine, originally known as the oil engine. His first prototype, built in 1893, will explode. But a working version constructed by 1897 will reach 75-percent efficiency, compared to 10-percent efficiency for the steam engine.

1894 After two unsuccessful efforts by members of the Geographical Survey of Canada to drill for gas near the Athabasca River, they move locations to a site near the Pelican River. At 250 metres the drillers strike an enormous reservoir of natural gas, which will blow wild for twenty-one years.

1898 New York City generates energy from incinerated garbage.

1901 Large oil strikes like the one at Spindletop, Texas, by Captain Anthony F. Lucas and his famous "Lucas Gusher" lead to the emergence of more than 600 new oil companies. Although many of them will not survive, several will grow into the giants of the energy business, including Sun, Gulf, Shell, and Texaco.

1902 Willis Havilland Carrier realizes that air can be dried by saturating it with chilled water to induce condensation. He builds the prototype for the Sackett-Wilhelm Lithographing and Publishing Company in Brooklyn. The device is used to help the printer combat the effects of heat and humidity on its coloured inks. Carrier will patent his "Aparatus for Treating Air" in 1906, and only later is the term air conditioner applied to his new invention.

1903 The Wright brothers put an internal combustion engine in an airplane, and on December 13, near Kitty Hawk, North Carolina, the *Wright Flyer* flies for twelve seconds over a distance of about 37 metres to become the first powered airplane to fly with a pilot aboard. Marie Curie becomes the first woman to win a Nobel Prize in physics, for helping to discover radium.

1905 Twenty-six-year-old Albert Einstein publishes his equation for the nature of energy ($E = mc^2$). His three-page paper does not immediately attract significant discussion, but will eventually lead to commercial nuclear electric power.

▼ CONTINUED ON PAGE 238

09.

REVERSE ENGINEERING: SOFT ENERGY PATHS

SUSAN HOLTZ, DAVID B. BROOKS

EDITORS' INTRODUCTION

U.S. President Jimmy Carter had spent the last ten days in apparent paralysis, trying to grapple with the energy crisis that was consuming the country. Only four months had passed since the partial core meltdown at the Three Mile Island nuclear plant, which had caused not only radiation but panic to fill the air. Meanwhile, events in the Middle East were once again affecting the American consumer. With the U.S.-friendly Shah of Iran replaced as national leader by Ayatollah Khomeini, OPEC suddenly raised the price of oil by 50 percent. Once again, Americans faced lineups at the gas pumps and a depressed economy, much like they'd just been through in 1973–74.

At Camp David, the president received a steady stream of experts and advisors all attempting to fashion an effective energy policy that might alleviate the situation. The problem was obvious: the United States was too dependent on foreign oil, and so its economy was in the grip of countries that wished to do the U.S. harm. The solution was less obvious.

On July 15, 1979, in what has been dubbed the "Malaise Speech," Carter went on television and spoke to the country. It was no mere policy speech; rather, the born-again Southern Baptist president probed the modern American soul. He spoke of a spiritual malaise that was gripping the United States. And, as Carter saw it, energy was at the heart of this malaise. "Energy will be the immediate test of our ability to unite this nation," Carter said. "It can also be the standard around which we rally. On the battlefield of energy we can win for our nation a new confidence and we can seize control again for our common destiny. . . . So, the solution of our energy crisis can also help us to conquer the crisis of the spirit in our country."

His six-point plan to confront the energy crisis was radical and far-reaching. It included the promise to develop local energy sources such as coal and solar energy. It included enormous financial support for energy conservation. And it contained as well a promise: "I am tonight setting a clear goal for the energy policy of the United States," the president pledged. "Beginning this moment, this nation will never use more foreign oil than we did in 1977 — never."

The OPEC-orchestrated energy crisis, along with the capture of the American hostages a few months later, destroyed Carter's presidency, but it also gave rise to a fundamental reconsidering of energy and how we use it. For perhaps the first time, energy became overtly entangled with fundamental questions of democracy, self-sufficiency, and independence. It became intertwined with the values of a society.

It was no accident that at the same time, a new approach to energy planning was born. Called the "soft energy path" approach, it set to deal with the complex set of values that were the underpinning of how any society considers — and plans for — its energy future. The soft path model gave us a still-useful tool for understanding how energy affects the kind of society we want to live in. The soft path approach has become a mature and powerful way of seeing our energy future and shaping our present. Susan Holtz and David B. Brooks, two of Canada's foremost soft energy path analysts, tell that story.

INTRODUCTION:
WHY THIS CHAPTER GOES AGAINST THE WHOLE CONCEPT OF THIS BOOK

Let's think back to the early 1970s, when this story begins.

The modern environmental movement had by then become a political force, albeit one that was far from having broad acceptance in the mainstream. The developing concern for environmental quality had already spawned new legislation and government agencies and, most significantly in this context, a multitude of local, national, and international citizen-based environmental groups.

The 1970s saw an ever-growing number of proposals for new energy megaprojects coming under consideration, and many people — mostly environmentalists, but others as well — in Canada, the United States, and around the world began turning their attention to issues raised by such unprecedentedly large projects. These issues were as diverse as the projects, which ranged from hydro dams in wilderness areas to oil and gas production in the Arctic and the offshore, to tar-sands plants, to oil tanker and liquefied natural gas terminals, to nuclear generating stations and the new uranium mines and refineries that supply them. Given the potentially huge impacts of these projects, social and economic as well as biophysical, environmentalists began to raise questions about the actual need for so much energy. Could there really be a demand for some thirty nuclear reactor units in the Canadian Maritime provinces by 2000, as some Atomic Energy of Canada Ltd. projections suggested?

For the two preceding decades energy demand had indeed been growing rapidly, fueled by low oil prices and the post-war economic expansion. For many of the megaproject proponents, the link between growth in energy use and growth in the GDP seemed not only strong historically but inflexible and essential. Of course new supply projects would be needed!

Then in 1973 came the first oil shock. With little warning the Arab oil producers embargoed the United States for some months, and the Organization of Petroleum Exporting Countries (OPEC) declared they would no longer negotiate but would instead adopt a take-it-or-leave-it approach to oil pricing. By the end of 1974 the price of a barrel of oil was eight times higher than it had been five years earlier.[1]

The resulting impacts shocked the Western world, both economically and politically. Higher home heating bills, long lineups at the gas pump, and higher costs for transportation-dependent goods and services (even for electricity, where utilities used oil-fired thermal plants) brought the situation directly home to consumers. Politicians and business leaders overnight became concerned about the security of the energy supply, and many called on governments to accelerate plans for new domestic energy sources. And then, toward the end of 1978 and into 1979, tight oil supplies and political events in the Middle East brought a second series of major oil-price hikes, which added to the political and economic consternation.

The stage was thus set after 1973 for the fierce battles that erupted between environmentalists opposed to new megaprojects and mainstream interests who insisted on the crucial need to increase domestic energy supplies. The intensity of some of these conflicts was not lessened because both sides had valid points to make. Not only did these massive projects entail many immediate social and environmental costs, an ecological perspective shone a hard new scientific light on other impacts that were subtle and distant in time and space. Discoveries such as the fact that trees were being damaged by acid rain hundreds of kilometres distant from the fossil fuel sources of the emissions, and that lead in gasoline might affect the behaviour and IQ of children heavily exposed to automobile exhaust, were reported in the media almost weekly. At the same time, it was clearer than ever before that the very fabric of society was dependent on energy, and that problems with energy supply and sudden price jumps did pose real threats to society's stability and well-being.

Out of this intellectual impasse the concept of soft energy paths was born. The term is taken from analyst Amory B. Lovins's first major essay on the subject, "Energy Strategy: The Road Not Taken?" which was published in the prestigious journal *Foreign Affairs* in the fall of 1976. Analytically, Lovins drew two different,

ANALYSTS DOING SOFT PATH STUDIES PROVIDED THE COMPREHENSIVE OVERVIEW, INCLUDING THE HARD NUMBERS DETAILING ENERGY SUPPLY-AND-DEMAND BALANCES WITH ALTERNATIVE TECHNOLOGIES, THAT WAS REQUIRED FOR A SERIOUS RESPONSE.

and opposing, pictures of energy policy, one relying on centralized, large-scale, capital-intensive technologies to meet rapidly rising demand (the "hard path"), and the other, the "soft path," emphasizing energy conservation[2] as the primary focus, with mainly smaller-scale renewable energy supply options selected to be more environmentally and socially benign.

We will discuss the content, the analytics, and some results of the soft path approach later in this chapter; for now, however, the key point is that Lovins (and subsequent soft path analysts) took both sides of the energy security vs. environment debate absolutely seriously.

Environmentalists taking a soft path approach to energy policy accepted that decision makers must consider and plan for energy security, and that economics did matter. Moreover, they met head-on one of the main criticisms of their opposition to various megaprojects. Instead of merely pointing generally to more efficient technologies and renewable energy sources when challenged as to how, exactly, the demand for energy was going to be met, analysts doing soft path studies provided the comprehensive overview, including the hard numbers detailing energy supply-and-demand balances with alternative technologies, that was required for a serious response.

Never before — and rarely since — had environmentalists developed a hard-headed vision for an entire sector of the economy. Analytically, it meant designing rigorous scenarios that took into account real-world constraints, and making choices for those scenarios that were not always ideal, yet kept sight of over-riding environmental and socio-economic goals. Psychologically, it meant behaving like a decision maker rather than a critic; some might say it meant behaving like a grown-up. For social critics, this is not the most comfortable stance, nor is it necessarily the most effective way to stop some specific project that you oppose. But it provides a model for public policy debate that is especially valuable in the more recent context of sustainable development, and almost essential for environment-related issues where there are a multitude of problem sources and varying consequences of dealing with them, and where there is no one best route to achieving economic, social, and environmental goals.

So how does a soft energy path approach differ from the concept of this book? In three main ways:

First, how to "fuel the future" — that is, focusing on the supply side of the energy equation — is *not* the soft path's primary question. Rather, the soft path asks first what kinds of energy services are needed for a given society at some future point. We will elaborate the point about "services" later. For the moment, it is only necessary to emphasize that energy demand reduction through different kinds of efficiency improvements and alternative choices is *always* the first priority.

Second, and even more fundamental, soft energy paths are not primarily about technologies, or even about human ingenuity in inventing technologies. They are first and foremost about *values*. A soft energy strategy is driven by a set of choices that explicitly take into account key environmental, economic, and social considerations. The choices about which considerations to incorporate are profoundly normative; a different set or weighting of these values-based choices would mean a different outcome for the strategy. Of course, no approach to energy policy is value-free. However, most policy analysts treat values implicitly by burying them in the analysis. Soft path analysts, in contrast, not only make the values explicit but also insist that they, along with the values-related implications that are de facto results of all energy decisions, should be the starting point for analysis.

Third, a soft path approach relies on a soft path *strategy* — choices based on an analytically rigorous *overview* of a specific society's energy demand and supply in the future. This approach contrasts with the way energy policy is most often presented in the media — as a patchwork constructed by various enthusiasts, each with a vision of a grand future for some particular technology, whether solar energy, hydrogen, nuclear fission, nuclear fusion, or whatever. True, it's good to understand as much as possible about what various technologies might achieve, and there is a valuable role for single-minded innovators and promoters of technology. But looking at energy futures through their eyes alone tends to give a perspective that is compelling but oversimplified and incomplete.

The soft path approach stands back from particular technologies billed as "the answer" and evaluates their usefulness in terms of the overall picture of the entire energy sector. Among other things, this approach includes looking at what, when, and

ALL TECHNOLOGIES, AND, INDEED, ALL PURPOSIVE HUMAN ACTIVITIES, HAVE NOT ONLY THEIR INTENDED RESULTS, LIKE PROVIDING ELECTRICITY OR PERSONAL TRANSPORTATION, BUT ALSO A HOST OF UNINTENDED EFFECTS, SUCH AS AIR POLLUTION, TRAFFIC CONGESTION, AND VULNERABILITY TO ICE STORMS AND TERRORIST ATTACKS.

where specific energy services are needed. It also means reviewing cost, timing, and political, social, and environmental implications as comprehensively as possible. It includes using some "less bad" fossil fuels, notably natural gas, that are needed as transitional sources on the way to less damaging alternatives. The ecological dictum that "you can never do only one thing" is its touchstone. All technologies, and, indeed, all purposive human activities, have not only their intended results, like providing electricity or personal transportation, but also a host of unintended effects, such as air pollution, traffic congestion, and vulnerability to ice storms and terrorist attacks. Energy efficiency itself can have some negative implications, for example if tighter houses don't have adequate ventilation. Thus a soft path strategy, even while making difficult choices, keeps as comprehensive a view as possible of context, constraints, and a full range of implications, recognizing that there are no simple, all-encompassing solutions.

A CLOSER LOOK AT THE SOFT PATH

To start with, it's worth making distinctions among a *soft path approach*, *soft technologies*, and a *soft path strategy* or *study*. To take the latter term first, a soft path strategy or study refers to a comprehensive demand/supply analysis using a soft path approach and employing, as much as possible, soft technologies. We will review its distinctive analytical steps in a later section.

At its basic level, a soft path approach, the most general term, is concerned with values. A soft path approach to energy policy focuses on reducing the risks of environmental and social harms that can result from energy sources and energy-using technologies, but it also accepts the importance of social and economic stability, full employment, and democratic institutions and norms. In the latter set of values the soft path approach does not differ markedly from most conventional energy planning in the Western world. (Energy planners of all stripes do tend to take economics and security seriously, even when their specific projections turn out to be dead wrong.) But the willingness to rule out altogether certain energy options on social and environmental grounds, and, for those same reasons, to emphasize the essential need for demand reduction and various mainly renewable technologies marked the soft path approach as a dramatic and controversial break with conventional policy twenty-five years ago.

Over the years, different soft path analysts have formulated the key characteristics of soft technologies — technologies appropriate to a soft path approach — in various ways. The key points behind the different formulations, however, are the same:

- Minimizing energy demand is far and away the most effective strategy for the environment;
- Renewable energy sources address sustainability and thus security over the long term; and

- Scale and diversity criteria are concerned with society's social and economic adaptability and resilience.

Some analysts add such terms as *relatively environmentally benign* and *safe-fail* to underline the safety and environmental concerns driving the analysis, or *decentralized* to make that element more explicit. But the classic description is the following one, taken from Lovins, who cites five characteristics of soft technologies:

- They rely on renewable energy flows that are always there whether we use them or not, such as sun and wind and vegetation: on energy income, not on depletable energy capital.
- They are diverse, so that as a national treasury runs on many small tax contributions, so national energy supply is an aggregate of very many individually modest contributions, each designed for maximum effectiveness in particular circumstances.
- They are flexible and relatively low technology — which does not mean unsophisticated, but rather, easy to understand and use without esoteric skills. . . .
- They are matched in scale and in geographic distribution to end use needs, taking advantage of the free distribution of most natural energy flows.
- They are matched in *energy quality* to end use needs. . . .[3]

The first three characteristics are easy enough to understand, but the last two, and especially the terms *end-use needs* and *energy quality*, require some explanation.

Much planning around energy is concerned with the availability and price of primary energy supplies such as coal, crude oil, and natural gas. However, what we really want is not energy for its own sake — having a barrel of crude on the front porch, or a pile of coal in the backyard — but for the *energy services* it provides. We don't want natural gas for our furnace, we want a warm house; we don't want gasoline for the car, we want to be able to haul things and get around. For analytical purposes, all these *end-use energy needs* can be grouped into the categories of lower-temperature heat, such as for warming homes or water; higher-temperature heat needed for industrial purposes; electricity-specific needs, such as for electric motors, lighting, and electronics; and motive power needed for transportation. (Transportation requirements have usually been referred to in terms of fluid fuels, since transportation requires carrying the source of motive power around with you and the high energy density of fluids makes that convenient. Nevertheless, technological advances in batteries and hydrogen allow for some other possibilities.)

Differences in energy quality were known to physicists for many years, but they were introduced into energy policy by soft path analysts. Our energy accounting system measures energy (the ability to do work) in common quantitative units like the kilowatt-hour, the joule in the International System of Units (SI), or the British thermal unit (Btu). In this sense, a joule is always a joule in that it always performs the same quantity of work. However, from a user's point of view, the kind of work (energy service) that you need is what matters. Different forms of energy, like electricity or solar radiation or natural gas, have different capabilities for delivering these different categories of energy services. The major Canadian soft energy path study conducted by Brooks et al. describes this concept this way:

> The quality of an energy form is a measure of the amount of "useful work" that can be extracted from the total energy, or enthalpy, contained in that form. For example, one joule of electricity can be used for many more things than a joule of low temperature heat, although the quantity

of energy is the same in both cases. [That is, electricity can be used for low-temperature heat through baseboard coils, for the high-temperature heat of kilns, for motive power, and for lighting and electronics.] The joule of electricity is a *higher quality form of energy* because it can provide more useful work than the joule of low temperature heat.[4]

Using technologies that match energy quality to end-use needs has relevance to a soft path approach and the development of soft path strategies for several reasons. A large portion of society's energy needs are for lower-quality energy in the form of low-temperature heat. (Think about raising ambient room temperatures to comfortable levels, a matter of tens of degrees at most.) Using a high-quality form of energy, such as electricity, to provide these services reduces efficiency in terms of its thermodynamic potential. It represents a loss in "second law" efficiency, so-called because it is based on the second law of thermodynamics. (As phrased to many students of physics to help them remember, the first law of thermodynamics states, "You can't win: energy can't be created or destroyed, only changed in form." The second law states, "You can't even break even: any transformation of energy degrades it, making it less organized and less able to do useful work.") Traditional or "first law" efficiency is about improving the ratio of the energy input to the useful work output of some device like a furnace or a motor. Second law efficiency is about using low-grade energy for low-grade needs, avoiding the use of high-quality forms such as electricity for those same low-temperature requirements. It also requires that the vast quantities of energy that are commonly treated as waste heat be used. (Increasingly now, the two-thirds of energy that is lost when fossil fuels are burned to generate electricity is being recaptured and used for industrial purposes or space heating.)

First law efficiency has always been recognized as important because it is obvious that it can conserve resources and save dollars. Soft path analysts, however, were among the first to see that an analysis of energy demand that categorized the quality of energy needed could further increase the potential for energy conservation. They recognized that conventional planning for future energy needs was based on past trends that relied heavily on high-quality conventional supply sources for low-grade end uses and included these large losses of useful energy through conversion. This loss, in turn, pushed up the projected demand for electricity generation and primary energy sources like crude oil.

Paying attention to this qualitative dimension of energy use also provided a better understanding of how renewable energy resources could be used more efficiently, adding scope and direction to their potential. And a focus on energy quality and on the tasks for which energy is used permitted detailed bottom-up analysis of a wide range of alternatives, including lifestyle choices and different sets of economic activities. This kind of analysis is the essence of the scenario-based soft energy path studies that were developed around the world.

SO HOW DO YOU DO A SOFT ENERGY PATH STUDY?

We assume that readers don't actually want to do a soft energy path study themselves, and so we will not present a detailed discussion of methodology here. However, it is useful to understand something about how to undertake a soft path analysis in order to better grasp what it can tell us — and what it cannot.

The analysis has five basic steps. Before describing them, though, it's important to note that the purpose of a soft energy path strategy is *not* to forecast the future. Rather, it's to explore the viability of some particular future scenario that reflects an emphasis on certain environmental, social, and economic considerations.

Step One — Building Scenarios

The first step is to determine the study's time frame and to create scenarios for what the society could look like at that future point. Major changes in the energy sector take a long time to happen. A saying in the utility business is, "Ten years down the road is yesterday," meaning that a large new generation facility typically takes at least ten years from the decision to build through financing and regulatory approval to completion. Cars and major appliances such as furnaces and hot water heaters take a decade or longer until they need replacement, and housing stock, if they are well maintained, can go on almost indefinitely. Thus, to allow for major changes in direction to have an effect, analysts must choose a date some decades in the future. For example, the large Canadian study by Brooks et al., published in 1983, set 2025 as its end point, with an interim analysis for 2000. (Of course, the further in the future, the more scope for change, but the fuzzier the details.) The researchers must then build up a picture of the whole society, determining a myriad of factors that affect energy use, such as population, household size, economic growth rates, and the industrial mix. The scenario must be tied together so that, for instance, enough shingles and steel and lumber are produced to build the projected housing required (something that can be done using an input-output model of that society's economy). Usually, more than one scenario is developed, showing a range of possible futures incorporating smaller and larger degrees of lifestyle change, economic growth, and so forth.

Step Two — End-Use Demand

Next the researchers must figure out the minimum quantity of energy required to run that society. This involves categorizing all the energy-using tasks and activities in this future society by end use, that is, by the quality of energy needed, and quantifying those categories. This step is organized by major sector; typically, households, commercial/institutional, industrial, and transportation. Each sector is usually subdivided and analyzed according to its particulars; for example, in the household sector, types of housing stock, housing stock turnover, household size, hot water use, appliance turnover, and new appliance penetration rates must all be determined. New energy-saving technologies and practices are incorporated in the study, but only as they become reasonably economical and practical, which also must be assessed.

Step Three — Supply Mix

This step involves matching end-use demand with supply options, again employing soft technologies based on renewable sources as they become reasonably competitive and available, and using fossil fuels and other conventional sources as transition technologies. Some existing or potential options, notably new nuclear energy units and offshore Arctic oil (though not natural gas, which poses far less threat to the environment), are excluded in virtually all studies. However, analysts may also exclude specific local options for environmental and social reasons, such as dams on wild rivers or wind-farm sites that affect scenic views. The idea is to test the degree to which soft supply technologies can, in fact, meet demand in the future, taking into account availability, technical considerations, and cost. Availability refers not only to obvious factors like whether any hydro sites are left to dam, but also to competing demands for the resource, such as sawlogs and fibre for wood, or soil amendment for crop waste. Technical considerations involve such things as the fact that the infrastructure to support major changes in transportation fuels must be co-ordinated and widely available, or that ensuring the stability of an electrical grid limits the percentage of intermittent sources (such as solar and wind) to about 25 percent.

Step Four — Backcasting

Backcasting tests the feasibility of "getting there from here." (The term was deliberately chosen to distinguish the step from forecasting, or predicting the future.) It involves a careful review of the technical and economic path that is needed to get from the present to the future scenario(s), a process that usually involves several iterations. In each iteration, various technical and other possibilities are identified, and the demand and supply analyses are revisited to consider alternative choices.

Step Five — Implementation

Because the purpose of a soft path strategy is to describe an alternative energy future, it is not surprising that the final step is to identify the conditions and interventions that are needed to bring about that scenario. High conventional energy prices were an important driver for the results obtained in most soft path studies; the stability of prices in recent years explains, in part, the failure to move farther along the soft energy path than we have. We will return to implementation in the last section of this essay. Here, it is only important to remember that the goal is not just to produce a scenario but to identify the policies and programs needed if it is to be achieved.

TYPICAL SOFT PATH RESULTS AND FURTHER DEVELOPMENTS

The Canadian soft energy path study, which appeared in book form at the height of soft path interest (Bott et al, 1983), was one of the largest and most detailed ever done, but it was by no means the only one. By the time it was completed, some thirty-five other soft path studies had been published, for countries ranging from Denmark to India.[5]

The more conservative of the two scenarios of the Canadian study concluded the following:

Under conditions of strong economic growth (an increase of more than 200% in GDP) and moderate population growth (an increase of over 50%), it would be technically feasible and cost-effective to operate the Canadian economy in 2025 with 12% less energy than it requires today, and over the same 47-year period, to shift from 16% reliance on renewable sources to 77%.[6]

It is worth noting that the study results for the same scenario at the interim point of 2000 showed an *increase* in energy use of about 4 percent. This simply demonstrates how long it takes for major improvements in energy efficiency to take hold as a result of the slow turnover of old capital stock.

The dramatic Canadian results were not atypical for soft path studies. For example, a Harvard Business School report on energy futures cited a study done in 1978 by a panel of experts

COMPARED WITH CONVENTIONAL FORECASTS, THE EARLY SOFT PATH STUDIES WERE OFTEN SO STARTLING IN THE POTENTIAL THEY IDENTIFIED FOR REDUCTIONS IN ENERGY DEMAND THAT THEY WERE DISMISSED OUT OF HAND.

from the National Academy of Science in the United States. One scenario for 2010 demonstrated that "very similar conditions of habitat, transportation, and other amenities could be provided in the United States using almost 20% less energy than used today."[7]

Compared with conventional forecasts, the early soft path studies were often so startling in the potential they identified for reductions in energy demand that they were dismissed out of hand. However, as the effects of energy conservation began to take hold in the 1980s, these results seemed less ridiculous. Some of the analytical features that were innovative in the first soft path strategies done in the 1970s, notably the bottom-up demand analyses, began to be incorporated into mainstream energy planning. More generally, the kind of thinking that started with environmental and socio-economic concerns and used a systems approach and tools like backcasting, scenario development, and analysis of end use began to be applied to more limited areas related to energy issues such as greenhouse gas emissions and to other fields like solid waste management and, especially, water management.[8]

THE ENERGY WORLD POST-1985: ARE WE THERE YET?

Energy prices stopped climbing after the mid-1980s, and, in the case of crude oil, even dropped back nearly to 1973 levels (a drop caused largely by the energy glut brought about by reduced demand). Prices then remained relatively stable until after 2000, when electricity and natural gas prices started to rise. Efficiency improvements continued to gain some momentum, based on the fact that their scope was enormous and many made excellent economic sense even without further energy price hikes. Better information and government programs, as well as wider awareness of the multiple benefits of energy efficiency, also played strong roles. However, in light of lower-than-anticipated conventional energy prices, the analysts who did the Canadian soft path study had to revisit their earlier scenarios. In a shorter, revised study released in 1988,[9] Torrie et al. incorporated oil prices that were roughly half those used earlier.[10] Their conclusions about the viability of a much more efficient, conservation-oriented future remained nearly the same, with the scope for energy savings only somewhat more modest. However, the competitiveness and, given the immaturity of the technology and other barriers, the practicality of most renewable options were significantly reduced. Their percentage of the hypothesized energy mix was much smaller, with wood and hydroelectricity the only economically attractive renewable sources. The other renewables had been replaced mainly by natural gas.

As had been the case in the earlier study, the most problematic area of the supply side was transportation fuels. The 1988 analysis dropped methanol as the proposed longer-term transportation option because of more information and concerns about herbicides and soil nutrient depletion in dedicated plantations for energy using fast-growing poplar, but allowed a greater role for ethanol. Nevertheless, all the transportation options raised some concerns.

During the 1990s, the return to a "normal" world energy situation lessened the political and public interest in energy issues, especially security of supply. (That particular topic, of course, has drawn more notice since the September 11, 2001, terror attacks.) As well, moderated demand weakened the viability of many proposed megaprojects and the Chernobyl accident specifically decreased acceptance of nuclear energy. Environmental attention related to energy began to centre on greenhouse gas emissions and air quality.

How does today's world compare with that envisioned in the original soft path studies? Each country is different, of course, and the differences between the challenges faced by developing countries and those of the developed world are great. Canada, as a typical developed country but with its own unique circumstances (e.g., cold temperatures, small human population in a gigantic land mass, abundant resources), is a reasonable example. Since the major soft path study done for Canada has 2025 as its end point, perhaps the right question is not "Are we there yet?" but rather "Are we going in the right direction?"

Energy Conservation (Demand Reduction, Efficiency Improvements)

Compared with conventional forecasts from the 1970s, we are much closer to a soft path than to business as usual as it was then envisioned. For example, in the more conservative scenario of the 1983 Canadian soft path study, primary energy consumption in Canada for the year 2000 was backcast to be 10,115 petajoules; in actuality it was about 10,500 petajoules — 40 percent lower than the level projected in a 1978 study done in the Long-Term Energy Assessment Program of Energy, Mines and Resources Canada. There is now an immense range of new and improved technologies and methods to conserve energy. Some of them, like the use of waste heat for co-generation in industry, are nearly invisible to most people but valued by the businesses who save money by using them. Others, such as high-efficiency windows, furnaces, and building techniques, are equally unobtrusive as they slowly become part of society's capital stock. Virtually none of the efficiency gains comes from anything approaching self-denial and a hair-shirt lifestyle, as demand reduction was widely characterized in the 1970s. One worrisome trend, however, is that fuel efficiency gains in automobiles are being undermined by increased driving and the popularity of the new gas gluttons, the SUVs. Another striking difference between the 1970s and the present is that in the 1970s, exported energy was about one-fifth of domestic demand, whereas now it is around four-fifths and moving toward equality. Much of this energy production goes to the United States, where there has been little political commitment to energy conservation. Consequently, even if Canada were wholeheartedly to adopt a domestic soft energy approach, at this level of exports there would still be important adverse social and environmental impacts from the high rates of energy production in Canada and the high rates of energy use in the United States, where Canada, obviously, has no ability to influence energy demand or to address other aspects of energy and environmental policies.

Use of Renewable Sources

Except for hydroelectricity, most of today's sources for which were planned or in place by the 1980s, the adoption of renewables has been very slow in Canada. The only exception is wood, the use of which grew in some regions and in 1997 contributed twice as much energy as nuclear sources (6 percent vs. 3 percent).

This is not the case elsewhere in the world. Worldwatch Institute's 2003 edition of *State of the World* reports that installed wind capacity worldwide is nearly 25 gigawatts, about 70 percent of which is in Europe, and the European Union has adopted a goal of having 22 percent of its electricity generated from small-scale renewable sources by 2010. Wind and solar are now the world's fastest-growing energy sources.[11]

Hard Path Technologies

Other than those that were in planning or under construction at the time of the 1983 soft energy study for Canada, few new megaprojects have gone forward. To the contrary, several existing nuclear units in Ontario have been shut down, and although some are supposed to be restarted, it is not clear when, or even if, this will happen. Offshore natural gas production has gone forward off Nova Scotia, but this was foreseen and accepted as a transition fuel in the soft path. On the other hand, hydrocarbon exploration on the biologically productive Georges Bank was suspended under a ten-year moratorium, which was extended beyond its original 2000 end point for environmental and socio-economic reasons.

Overall, while Canada may not be exactly on the path mapped out, increased energy conservation and the curtailment of hard path projects indicate that it is not going in the opposite direction, either.

RE-VISIONING THE SOFT PATH PERSPECTIVE

Not surprisingly, soft path analysts think that their approach is, if anything, more appropriate now than it was twenty-five years ago. Since then, concerns about health and environmental problems related to energy have continued to grow, including avoidable respiratory illnesses and childhood asthma linked to air quality, and greenhouse gas emissions linked to fossil fuel use. It should be mentioned that soft path researchers were among the first to pay attention to climate change; Amory Lovins et al. published a book about soft path approaches in 1981 entitled *Least Cost Energy: Solving the CO$_2$ Problem*. Though not designed with greenhouse gas emissions in mind, even the higher energy use scenarios in the 1988 revision of the soft path study for Canada would meet our Kyoto commitments.

The strength of these early studies was recently confirmed by a study entitled *Kyoto and Beyond* that Ralph Torrie and colleagues prepared for the David Suzuki Foundation and the Climate Action Network Canada. Their sector-by-sector scenarios showed how readily available, cost-effective demand-side technologies

THE SOFT PATH ANALYSIS IS ALSO ESPECIALLY RELEVANT TODAY IN ITS EMPHASIS ON THE RELATIONSHIP BETWEEN A SOCIETY'S ESSENTIAL TECHNOLOGIES AND ITS RESILIENCE AND ADAPTABILITY.

could cut energy use by so much that Canada would not only exceed its Kyoto commitments, but consumers would save tens of billions of dollars per year.[12] The same is true elsewhere, as examples of "Factor 4" improvements — that is, reductions of 75 percent in resource use per unit of output in industrial and commercial activities — demonstrate.[13]

If nothing else, these scenarios show how much can be achieved, and without much pain — the latest scenarios have even allowed for a 50 percent increase in per-capita GDP — by following soft path prescriptions. However, they also show that pervasive energy-related problems inhere so much in the fabric of society that they cannot be addressed, let alone solved, without rigorous analytical work.

The soft path analysis is also especially relevant today in its emphasis on the relationship between a society's essential technologies and its resilience and adaptability. Again, Amory Lovins and his colleagues were far-seeing in their 1982 book *Brittle Power: Energy Strategy for National Security*, in which they identified some of the ramifications of highly centralized energy systems. Today's globally interlaced economic systems, personal

FROM A GLOBAL PERSPECTIVE, THE DEVELOPING WORLD'S ASPIRATIONS FOR A REASONABLE LEVEL OF PROSPERITY WILL MEAN THAT THEIR ENERGY USE WILL GROW, AND, IN ORDER TO ALLOW THIS WITHOUT CROSSING FURTHER ECOLOGICAL THRESHOLDS, CONSTRAINING ENERGY USE IN THE DEVELOPED WORLD IS IMPERATIVE.

connections, and political interactions make for a world of many surprises, ranging from SARS and the West Nile virus to terrorist attacks. And concerns about the role of nuclear power plants as sources of fissionable material for nuclear weapons and their potential as targets of military or terror attacks are, sadly, more credible today than they were in previous years.

Finally, the continuing need for prioritizing energy conservation cannot be overstated. The environmental and social implications of unchecked demand growth are enormous, whether we are talking about livable cities, habitat loss, security of supply, or any number of other issues. From a global perspective, the developing world's aspirations for a reasonable level of prosperity will mean that their energy use will grow, and, in order to allow this without crossing further ecological thresholds, constraining energy use in the developed world is imperative. Renewable sources alone are not and cannot be the answer. Only major success in reducing demand will put us firmly on the soft path.

Was the original soft path approach and its strategies right about everything? Of course not. First, the emphasis placed on finding low-tech and easy-to-understand technologies, as well as the aversion to *any* technologies involving large-scale, centralized facilities, was overstated. Hydrogen technology, or even a high-efficiency gas furnace, is not really much simpler to understand than nuclear power, though a toppled wind turbine is certainly a lot easier and cheaper to fix than a nuclear unit that is "down." And even though the hard and soft paths have opposite assumptions and approaches, the real world has elements of both at the same time.

Second, the soft path strategies were too limited in their analysis of what drives change. Comparatively high prices for conventional fuels and rational economic behaviour were presumed to be the main motivators for change in the original studies. Though the right economic signals are essential, it is now apparent that other factors are equally or perhaps even more important. As Ralph Torrie put it, "The demand is for better services, not for more energy-efficient services. The fact that better buildings (and more profitable factories, better vehicles, more compact urban forms, etc.) are also more energy efficient is a good thing, but the benefits, including the financial benefits, are much greater from the way the service is improved than from the fuel and electricity savings they deliver."[14]

Third, some political realities that hold back change needed to be more effectively addressed. Governments, both politicians and the bureaucracy, always prefer a situation where there are no (visible) problems to one where they are expected to solve problems. Not surprisingly, then, the years of energy "normalcy" were unproductive for soft path initiatives, especially those that supported renewable technologies. This inertia was strengthened in Canada by program and staff reductions as governments struggled to get deficits under control, and in the United States by the election of governments unwilling to support pro-environment interventions for fear they would limit business. In addition, apart from times of energy crisis, few politicians worry much about how much energy is being used. Rather, they bias policies to protect industries in their constituencies, including energy supply industries. Revenue stability from taxes and royalties also matter: it's hard to believe that the Alberta government's desire to maintain its substantial revenues from the energy sector did not play a role in its opposition to the Kyoto Accord.

Finally, the concept of sustainable development has highlighted the need to consider equity implications of policies alongside economic and environmental matters. For example, incorporating environmental costs into energy prices would likely raise prices, and the effect of this on particular groups of consumers and of employees in the energy sector needs evaluation. Low income is not the only factor causing vulnerability. Rural location, which affects transportation, as well as age, education, family situation, or even being new immigrants are other factors that should be examined as creating the conditions for unfairly burdensome impacts.

The world today has also brought a number of changes that certainly affect what a soft path will look like, but in ways that are not yet clear. Among the most important are the deregulation and reorganization of utilities and the electricity market; the role of exports in countries like Canada that have or could build the capacity for dedicated energy export facilities; and how key emerging technologies will evolve, especially in the transportation sector, where change is also closely linked to changes in settlement patterns, urban and suburban planning, trade, and the structure of industry. But however many misty patches there are in a re-visioning of the soft path, its direction remains fixed by the lodestar values of sustainability — maintaining and improving human well-being and the economic and social structures that support it, and, equally, maintaining and restoring the ecosystems that surround and support human activities and of which we, too, are a part.

In summary, the general soft energy approach and the specific soft energy study for Canada have both proven to have lasting, and indeed growing, value. On the one hand, the Canadian study turned out to be more prescient than even the fifteen or so analysts who worked on it might have expected. On the other hand, both the approach and the study overestimated government's willingness to recognize the value and act on the principles of the soft path. Nowhere is this contradiction better illustrated than in the changing reactions of Canadian energy bureaucrats over the course of the study. In the early days of soft path analysis, the typical response was, "Yes, you have created a very desirable future for Canada, but do you *really* think it is feasible?" After the study was completed, the response changed: "Yes, you have shown that this future scenario is feasible, but do you really think it is *desirable*?"

A HISTORY OF ENERGY **(CONTINUED)**

1907   James Murray Spangler, a department store janitor in Canton, Ohio, invents the first electric vacuum cleaner by rigging a carpet sweeper with a fan motor and a soapbox.

1911   The U.S. government breaks up Standard Oil under the Sherman Antitrust Act. The first filling station in the United States opens in Detroit.

1913   Henry Ford installs the first moving assembly line at the Highland Park factory. This innovation allows him to manufacture 1,000 cars a day.

1914–18   Under the direction of Winston Churchill, the British fleet is converted from coal power to oil. This allows the British to achieve naval superiority over Germany in World War I and helps to establish oil as a key strategic commodity.

1928   Frank Whittle, a pilot for the RAF, begins his research on jet propulsion. His designs will lead to the maiden flight of a jet aircraft on May 15, 1941, ushering in the modern era of jet aviation.

1929   U.S. automobile manufacturing consumes 20 percent of the production of U.S. steel, 80 percent of its rubber, and 75 percent of its plate glass.

1930   Oil replaces coal as the world's main fuel for transportation.

1935   Following a conversation with a truck driver who lost a shipment of chickens because the storage compartment overheated, Frederick M. Jones invents the first automatic refrigeration system for long-haul trucks. His system, which will be patented in 1943, will later be converted for ships and railway cars, changing forever the American food industry.

1936   The Hoover dam is finally completed, after five years of construction. The largest dam of its time, it stands 221 metres tall, weighing more than 6.5 million tonnes. It is the first giant multi-purpose dam in the world, with average annual power generation capabilities of about 4 billion kilowatt-hours.

▼   CONTINUED ON PAGE 294

10.

THE NEW BOTTOM LINE: ENERGY AND CORPORATE INGENUITY

DAVID WHEELER, JANE THOMSON

EDITORS' INTRODUCTION

By the time CEO Jeff Skilling's face graced the cover of *BusinessWeek* magazine, on February 12, 2001, Enron could do no wrong. The company had been ranked by *Fortune* magazine as The Most Innovative Company in America for the fifth year in a row while growing to over US$100 billion in revenues. Fawned over by fund managers, politicians, and presidents, Skilling and his team of senior managers had used their power to transform energy markets in the United States like no other company since Standard Oil, while making themselves fabulously wealthy.

Within a year, of course, everything had changed. It started with news of "off-balance-sheet" debts owing to single-purpose companies that went by the name of Jedi, Checo, and other Star Wars characters. By the time the whole story leaked out, Enron's off-book liabilities had reached an estimated $690 million and the company declared bankruptcy. The stock that once sold for $90 a share was worth pennies. Over 13,000 employees lost their jobs.

Unfortunately, Enron has become emblematic of the worst clichés used to describe the big energy business, and big business in general: callous, duplicitous, self-serving, and dangerous. It's not surprising then that public perception about big industry is so low. In a 2002 Environics International poll of 36,000 people conducted in forty-seven countries, 59 percent of respondents said they believed that non-governmental organizations like Greenpeace operate in the best interests of society, while only 39 percent believed that multinational companies do the same thing. In other words, much of the public believes that big business is part of the problem, not part of the solution. But that score might also reveal a lack of awareness of the dramatic transformation that's happening in the business world.

While it's a far less dramatic story than the collapse of Enron, the innovation going on in key parts of the energy industry is perhaps more significant, and might change some perceptions about how the big energy players operate. Gone, for instance, are the days when energy companies universally dispute global warming. Today companies like BP, Suncor, and Dofasco are actually leading the way in thinking through new approaches to energy sustainability. In the new model, environmental and social benefits are as fundamental to the bottom line as profits are.

David Wheeler and Jane Thomson have conducted an extensive survey of senior management of large energy-intensive industries, trying to track the arc of this new energy revolution. They answer the question so many of us want to know: Is the business community responding in a serious way to the issue of climate change, and is it preparing to meet society's growing needs for sustainable energy sources in the twenty-first century? Wheeler and Thomson investigate innovations in products and processes and the development of entirely new markets for sustainable power, heating, cooling, lighting, and mobility. They also draw attention to some of the constraints and uncertainties that businesses face from the political, economic, and social points of view in making the transition from unsustainable to sustainable energy use and distribution.

INTRODUCTION: THE RULES OF THE GAME ARE CHANGING

Since the dawn of the industrial era, business and industry have been both the providers and the principal users of primary energy resources. And the harnessing of technological processes in the extraction and use of these resources has had profound implications for society on a global scale ever since. Without the marriage of coal to the steam engine, courtesy of James Watt, there would have been no railroads or steamships and nineteenth-century empires based on international trade would have developed very differently. Without the development of oil products and their deployment in internal combustion engines, courtesy of entrepreneurs like Henry Ford, motorized personal transportation would not have emerged in quite the same way and many leading industrial economies would look entirely different today. The oil industry itself, driven from its earliest days by the zeal of business people like John D. Rockefeller and his successors, has for one hundred years exemplified the raw application of engineering to nature in pursuit of power, influence, and wealth.

These industrial developments and the social costs that accompanied them all had their vociferous critics.[1] From the early trade unionists concerned about industrial safety in coal mines, to the regulators and consumer activists that eventually broke the monopoly power of the oil magnates, to the social and environmental activists and anti-globalization protesters of today, energy-intensive businesses have always had to confront their critics head-on. Meanwhile, it may be noted, these debates have not arrested our insatiable desire for the many essential goods and services associated with the energy-intensive industries: transportation, heat, light, and infrastructure.

It is against the background of this "realpolitik" of societal choices that we address the possibility that business and industry might play an equally powerful but more socially and environmentally sustainable role in fueling the future. What are the prospects for a new enterprise-led energy revolution that will supersede all previous revolutions in terms of efficiency and technological sophistication and thereby deliver the safer and more secure energy future that ordinary citizens and policy-makers are demanding with increasing urgency?

Our perspective on this question is essentially an optimistic one. Part of the reason for this optimism is simple logic: given the trends described elsewhere in this book, there really is no alternative to the emergence of a radically different energy future and it is axiomatic that business and industry will be mobilized accordingly, through both market "pull" and public policy "push." The required human ingenuity simply has to emerge, because the consequences of failure would be so unthinkable. More pragmatically, we have examined evidence from many sources, much of which we present in this chapter, we have interviewed thirty leading businesspeople in Canada and internationally, and we have concluded that business and industry already have the capabilities — and indeed many of the technologies — to play a full part in, and indeed perhaps lead, the kind of energy revolution envisaged by many commentators. The only real uncertainties concern timing and economics.

The final reason for optimism is that businesses and their leaders are slowly but surely changing their natures. Increasingly, our largest and most influential corporations are not led by the rogues and dinosaurs of popular mythology (although given recent events in the United States, ordinary people could be forgiven for drawing a straight line between the era of the robber barons and oil magnates and the era of the high-profile scams at Enron). Happily, just as in the nineteenth century, when there were a few enlightened businesspeople acting in society's interests as well as those of their investors, so today in the twenty-first century, we can point to large numbers of ethically run, socially responsible businesses in Europe, Asia, and North America. And many of these businesses and their leaders want to be part of a new energy future because (a) they believe it makes sense for society, and (b) they understand that is where the money will be. Let us now examine this phenomenon in more detail.

THE ROLE OF BUSINESS IN A SUSTAINABLE WORLD

Sheikh Ahmad Zaki Yamani, the former Saudi petroleum minister, is famously credited with asserting that the stone age did not come to an end because the world ran out of stones, and thus the oil age will end long before the world runs out of oil. Today, increasing numbers of influential commentators believe that the world will begin to reduce its reliance on fossil fuels and convert to more sustainable energy sources because it makes sense environmentally, socially, and *economically*. Inspired by this simple truth, Amory Lovins, one of the world's most eloquent technology-inspired environmentalists and a consummate advocate of new energy futures, has written powerfully on the potential for "natural capitalism."[2] Natural capitalism is a new form of business endeavour that brings to life the notion of the "triple bottom line," in which business simultaneously builds economic, social, and environmental value for the firm and for society. As described by John Elkington, it is now possible for businesses to create simultaneous "wins" for their stakeholders and society at large, for their corporate profits and the environment.[3]

In support of these grand ideas, a flood of popular and academic literature has emerged in the last decade — often authored by leading business personalities — describing how business can be more environmentally conscious, socially engaged, and economically productive.[4] Supported by a burgeoning "business case" for sustainability, in the ten years since the 1992 Rio Earth Summit on Environment and Development, a number of significant changes have occurred in how business engages and is seen to engage with a range of pressing global problems — including climate change.

For example, a decade ago it would have been hard to find many business leaders who would have disagreed with the traditional perspective that global environmental and social issues were solely a matter for governments to resolve and that, on the whole, civil-society organizations and social activists were a nuisance. Today, it is common for business leaders to acknowledge that such issues will be resolved only by partnerships involving governments and civil-society organizations. In an Environics International survey of 212 business leaders across fifty countries in 2002, 62 percent of respondents said they believed that "producing and using clean energy" was best achieved through a partnership of companies and governments, compared to 9 percent who felt this issue was mainly for governments and 29 percent who felt it was mainly for companies to resolve.[5]

Interestingly, business attitudes to risk and opportunity have also changed. Today, many leading businesses recognize that environmental and social issues are a source both of risk and of opportunity, and that effective leadership, innovation, and stakeholder inclusion tips the balance in favour of opportunity rather than risk or liability. According to the Environics data, 80 percent of the 212 business leaders polled said they thought sustainable development was an opportunity versus only 20 percent who perceived the issue solely in terms of costs.

Governments are also looking increasingly for partnership-based solutions involving business, especially in public policy fields as complex as industrial policy — and energy supply and use in particular. David Bell of York University has described private research with G7 governments in 2002 that revealed almost no appetite for new regulation of the energy industry in the world's leading economies, but instead significant trends towards future application of voluntary agreements, market instruments and education, persuasion, and information provision for decision making.[6]

Despite these trends, business still faces a significant challenge. Corporations still lack public credibility on issues like climate change. Another Environics International poll conducted in 2002 explored attitudes of 36,000 citizens in forty-seven countries to the question of legitimacy of different actors to act in a socially responsible manner. The poll demonstrated continued relatively high trust in non-governmental organizations (NGOs) to operate in the best interests of society: 59 percent of respondents reported a lot or some trust versus 32 percent who said they had little or no trust in NGOs. This compared with only 50 percent who placed trust in government, 39 percent in global companies, and just 38 percent who placed trust in parliament/congress. Expanding on the challenge, in a presentation prepared for the Sustainable Enterprise Academy in May 2003, Doug Miller of Environics concluded from his research data that "problems are admitted and solutions are known — what's needed is delivery. . . . in any community, when challenges are great, all are expected to be part of the solution." Instead, what Miller sees is a perceived ineffectiveness of national governments, and the emergence of a "leadership vacuum."

So here is both the dilemma and the opportunity for business. Leading companies, governments, and many civil-society actors are developing the language and the new skills associated with multi-sectoral, partnership-based solutions to pressing global issues like climate change. And yet many ordinary citizens remain unconvinced by the intentions of the different actors — especially business. A major opportunity exists, therefore, for the three sectors, including business, to re-legitimate each other by demonstrating real leadership and real action. This will require some creativity and some risk taking. It may also require new institutional frameworks to help broker partnerships and allow different partners to learn their way into new ways of working. Happily, many new business organizations are creating precisely those opportunities.

The World Business Council on Sustainable Development (WBCSD) is a group of more than 160 of the world's leading corporations that are committed to the concept of sustainability. Interestingly, five of the seven Canadian members of WBCSD are in the energy business: three utilities (BC Hydro, Ontario Power Generation, and TransAlta) and two oil and gas companies (Petro-Canada and Suncor Energy); the other two members are heavy energy users (Alcan and mining giant Noranda). The WBCSD is CEO-led and has launched special projects in several energy-intensive sectors, including cement, electric utilities, mining and minerals, and mobility. It also has twelve cross-cutting themes and activities addressing such issues as accountability, risk, eco-efficiency (producing goods and services with less materials and energy), and energy and climate change.

Together with the International Chamber of Commerce, the WBCSD organized the business voice at the 2002 Johannesburg World Summit on Sustainable Development through an initiative titled Business Action for Sustainable Development (BASD). Under the joint chairmanship of Mark Moody-Stuart (former chairman of the Committee of Managing Directors at Shell) and Lord Richard Holme (of mining conglomerate Rio Tinto), BASD provided some of the most practical interventions at the event, including the creation of a Web site with many on-line resources relating to the summit (including numerous examples of partnership projects in sustainable development involving business). Other business organizations that have emerged in recent years include the United Nation's Global Compact grouping of companies (an initiative inspired by Secretary-General Kofi Annan), the International Business Leaders Forum (originally inspired by the Prince of Wales), Empresa (a corporate social responsibility grouping for the Americas), CSR Europe, U.S.-based Business for Social Responsibility and its Canadian counterpart CBSR, Business in the Community (U.K.), and the Imagine program of the Canadian Centre for Philanthropy. All of these organizations provide active fora for business leaders to meet and co-ordinate thinking on sustainable development and corporate citizenship.

We have also seen the emergence of powerful advocacy for sustainable enterprise by leading businesspeople in publications, speeches, and media appearances.[7] Perhaps the high point of this advocacy, co-ordinated by BASD, occurred on September 1, 2002 — the designated "business day" of the Johannesburg Summit. There the Secretary-General of the United Nations, Tony Blair, Jean Chrétien, and other political figures were joined by business leaders from a wide variety of sectors, including an especially strong contingent from the energy and related industries. Among the energy companies represented by senior business leaders were Shell (represented by Phil Watts, chairman of the Committee of Managing Directors), Eskom, the South African energy utility (Reuel Khoza, chairman), Electricité de France (François Roussely, chairman), Ontario Power Generation (Ron Osborne, CEO), and Toyota (Dr. Shoichiro Toyoda, honorary chairman). Such a lineup of senior business leaders on issues of global concern was wholly unprecedented — as indeed was the spectacle of Greenpeace and WBCSD holding a joint press conference and declaration on climate change at the same event.

The final declaration from BASD at the summit was unequivocal:

> We need to make sustainable development happen by generating economic growth with greater resource efficiency, minimizing environmental impacts, and with maximum social well-being for more people. We also welcome the growing realization that business is an indispensable part of the solution to the problems of the world. We have improved our relationships with governments, NGOs, and others. Together we will turn the idea of sustainable development through practical partnerships into a growing reality on the ground. As we move forward the view of business could be summarized in the words of Elvis Presley: "A little less conversation, a little more action."

Arguably, the energy industry has always been close to the leading edge on the business case for sustainability. Here we may cite the corporate re-invention of Shell following the searing experiences of that company in Europe and Nigeria in the mid-1990s and the seminal speech by BP's John Browne at Stanford University in May 1997, when he became the first major player in the energy industry to accept the implications of climate change, thereby splitting his industry down the middle. In Shell's case, two events happened in quick succession that demonstrated that their former business model was in need of significant overhaul. The Brent Spar incident occurred in summer 1995, when Shell UK obtained British government permission to dispose of an oil platform at sea. In response, Greenpeace organized major protests against the company across Europe which resulted in the rapid shelving of the disposal plan, thereby demonstrating that regardless of regulatory and technical opinion, activist perspectives and public reaction simply could not be bypassed. Later that year, environmental and social activist Ken Saro-Wiwa was executed by Shell's principal partner in Nigeria, the Nigerian military government. Again this incident brought Shell's strategy with respect to corporate responsibility and human rights under the glare of international media attention.[8]

Today, both Shell and BP have exemplary policies and practices on social and environmental issues; they are also vying for technological and commercial leadership in alternative and renewable energy, both having invested hundreds of millions of dollars in their non-oil-related businesses in recent years.

In Canada, we may cite the similarly visible leadership of Suncor Energy, a company that was set to be shut down by its U.S. parent in the early 1990s, but which, under the leadership of Rick George, has established an impressive reputation with respect to community relations, civil-society engagement, and environmental responsibility.[9] In addition, Suncor is a major investor in alternative and renewable energy development and was quick to accept with good grace the inevitability of Kyoto implementation despite its heavy investments in the carbon-intensive Alberta oil sands.

Such advocacy and high-level political engagement by the predecessors of these companies and their CEOs would have been unthinkable ten years previously. So what are the underlying influences that permit such engagement by business leaders and enable the shifts in corporate strategy that this engagement represents? At a governance level, why do investors and other stakeholders accept, and even encourage, such engagement? And why are such changes so important to the energy industry?

Strategic management theorists were the first to argue that competitive advantage could be gained from prompt and effective responses to social and environmental drivers in the economy. More than twenty years ago, guru of gurus Peter Drucker asserted that "the proper 'social responsibility' of business is to tame the dragon, that is to turn a social problem into economic opportunity and economic benefit, into productive capacity, into human competence, into well paid jobs, and into wealth."[10] Michael Porter, the Harvard University professor who towers over the field of strategic management, has developed this proposition, relating it directly to innovation. In his introduction to *Tomorrow's Markets*, Porter revisited his contention, first aired in the Harvard Business Review in 1995,[11] that environmental and social challenges can be a significant spur to industrial innovation. Porter asserts: "It is becoming more and more apparent . . . that treating broader social issues and corporate strategy as separate and distinct has long been unwise, never more so than today. . . . We are learning that the most effective way to address many of the world's most pressing problems is to mobilize the corporate sector where both companies and society can benefit."[12]

Henry Mintzberg, Canada's leading management scholar, has also been very active in his critique of outdated approaches to strategic management that suggest the application of rationalist economic nostrums in the absence of societal and social context. In an interview with a Scottish newspaper during the height of the unfolding U.S. corporate scandals, Mintzberg said, "The system is sick right down to its roots. Call it shareholder value or whatever, but the obsession with narrow performance and economics skews the way that people think and act. Businesses can't function without some degree of social responsibility."[13]

Rather helpfully, these leading management theorists move us beyond the old assumption that there has to be some kind of moral choice between doing good and being profitable. Others have developed this case further, making the point that the world is changing fundamentally as a result of economic and technological drivers and that, as a result, the way in which value is created today is entirely different than how it was created ten or twenty years ago.[14] The majority of firms' stock market value today is not based on "hard assets," for example balance sheet items, physical capital, and so on. Rather, it is based on the confidence of investors in the firm's ability to leverage "soft assets": its relationships with customers and suppliers (social capital) and its ability to recruit top talent (human capital), to innovate (intellectual capital), and to achieve positive brand identity (reputational capital). The brand development company Interbrand publishes annual ranking of brand valuations for leading companies. In 2002 they calculated the brand values of Coca Cola and Microsoft at US$69.6 billion and US$64.1 billion respectively; these are not insignificant assets. Even corporate culture (the values and beliefs of the organization) are now judged vital elements of competitive advantage in many sectors.[15] In this new world of soft assets driving value, it is unsurprising that business leaders — especially leaders of "high impact" industries like oil and gas, power, and automotive — have changed their strategies to reflect more of a "stakeholder approach," recognizing that these days what is good for customers, workers, suppliers, and local communities (within whose control almost all the soft assets reside) is usually good for sales, profits, and shareholder value too. This sort of strategy is the reason why BP got access to Alaskan oil reserves rather than its competitors and why Suncor was granted regulatory permissions on its multi-billion-dollar oil sands developments eighteen months early.[16]

The evidence of positive correlations between superior economic performance and superior social and environmental performance is now very strong, extending over a wide range of academic studies[17] and indeed from the practical evidence of "sustainable development" stock portfolios that typically

outperform all-share indices.[18] So although all these phe-
nomena are new and the evidence recent, it seems that today's
business leaders are on safe ground in asserting a case for social
and environmental issues driving economic performance rather
than diminishing it.

THE ASSUMPTIONS UNDER WHICH TODAY'S LEADING BUSINESSES OPERATE

In the field of strategic management, scholars have traditionally
classified key external drivers of business decisions as being
political, economic, social, or technological. Hence, researchers
and business consultants have called their assessments of the
external business environment "PEST" analyses. In today's world
we must also add ecological drivers as highly important, if not
crucial, to most industry sectors. Thus a PEST analysis might
now be more useful as a "PESTe" analysis if it is to be a complete
description of external factors impacting on business. Businesses
make assumptions about all of these drivers, and the best businesses try to predict where the drivers will go in order to avoid
being caught out by shifts in market conditions caused by changes
in the external environment.

Some businesses formalize their analyses of future trends
through rigorous and systematic application of "scenario planning" — playing with possibilities that might emerge through
political, economic, social, technological, and now ecological
change. One of the best-known examples of an international
firm employing scenario planning for the purpose of stimulating
internal and external learning, and thus enhanced strategic
thinking, is Royal Dutch/Shell.[19] Shell started scenario planning in 1968, when it established an ad hoc "Year 2000" study
group to address possibilities that oil might actually run out one
day. The group posed the question, "Is there life after oil?"
Subsequently, Shell's scenario planning team was generally

credited for postulating — well before the events occurred — developments such as the 1973 and 1979 energy crises, the growth in energy conservation, and even the breakup of the Soviet Union.[20]

So it is highly relevant that Royal Dutch/Shell has recently developed two new international energy scenarios to the year 2050.[21] One scenario, "Dynamics as Usual," envisages an early and dramatic expansion of natural gas use (especially in East Asia), a plateau ("boom and bust") for renewables by 2020, a minor but short-lived renaissance for nuclear energy, and the emergence of new, bio-fuel-based renewables by 2040. The alternative scenario — labelled "Spirit of the Coming Age" — envisages the emergence of a serious hydrogen economy in both static and transport-related applications (25 percent of all OECD vehicle propulsion is fuel-cell powered by 2025), the limitation of renewable energy to niche markets in early years, and the expansion of large-scale renewable and nuclear energy schemes from 2030 in order to produce the required hydrogen by electrolysis. In both scenarios fossil fuel continues to be the major source for meeting primary energy needs.

The key questions posed in Shell's research, and perhaps vital questions for anyone with an interest in what the world may look like in the future, were the following: when will oil and gas resources cease to meet rising demand, and what will replace oil in transport; which technology will win the race to improve the environmental standards of vehicles; how will demand for distributed power shape the energy system; who will drive the market growth and cost reduction of renewable energy sources; how will the problem of energy storage for intermittent renewables like solar and wind be solved; how might a hydrogen infrastructure develop; how will emerging economies like China and India balance rapidly growing energy needs with rising import dependence and environmental effects; and where will social and personal priorities lie and how will these affect energy choices?

Addressing these fundamental questions allowed Shell's scenario planners and expert advisers to integrate and build on powerful socio-economic trends — for example, the fact that global average per-capita incomes will be above US$20,000 by 2050, that 80 percent of the world's 8.5 billion population will be urban at that time, and that market liberalization and consumer attitudes will continue to promote cleaner and more efficient fuels. Thanks to some restraints because of greater efficiencies of energy production and distribution, Shell was also able to predict only a doubling of primary energy requirements by 2050 (rather than the trebling that some predict). Shell believes that at this time the world might start to approach "energy saturation" — a situation that arises because of evidence suggesting that historically, as per-capita annual incomes exceed US$20,000, annual primary energy needs flatten in all leading economies (albeit for E.U. and Japanese consumers at less than half that of the average U.S. consumer).

Whichever of Shell's scenarios emerges (and they are not mutually exclusive or guaranteed to happen), some common features are evident:

- Fossil fuels — oil for transport and gas for power — will remain the main source of energy for the coming fifty years;
- There is a general assumption that atmospheric carbon dioxide (CO_2) levels will likely rise to between 500 and 550 parts per million (ppm) before stabilizing and hence the world's economic, natural, and physical systems may simply have to adapt to these levels;[22]
- Natural gas is going to be an important bridge for the next twenty years under any circumstances;
- Oil markets will face disruption as new vehicle technologies diffuse — whichever these winning technologies are;

- Basic heat and power requirements will become more distributed and decentralized for economic and social reasons; and
- We are entering a period of technological innovation and uncertainty during which it is difficult to predict which technologies will win and therefore what will be the specific roles for fossil fuels, renewables, hydrogen, and nuclear power.

These observations are important to our analysis for two reasons: (a) they emphasize the importance of current business reality; and (b) they reinforce the importance of political, social, and economic (i.e., investment climate) reality in mediating technological change and market transformation. We will consider these two "realities" in turn.

Current Business Reality

Shell's scenario planning demonstrates one key point above all others. Even the most enlightened of mainstream businesses are not likely to plan their future strategies around any of the potential nightmare scenarios that are advanced by some activist groups and scientists — some of whom fear that atmospheric CO_2 levels of 500 ppm and above might be too much for the planet to accommodate.[23] Their fears include such catastrophic eventualities as melting polar ice caps, collapse of agricultural systems, malaria in the U.S. Midwest, and extreme variations in ocean currents and climatic systems. We should be clear here: these possibilities have not been entirely discounted in corporate scenario planning; it is simply that the best available evidence suggests that they are not the most likely outcomes and therefore they will not drive business behaviour — at least in the short term.

This means that we should not expect to see the public positions of the major oil companies and Greenpeace reconciled anytime soon. Business and activist groups are working from similar data but they have a different attitude to risk, and in particular the application of the "precautionary principle" — the notion that economic activity should not proceed in the absence of proven safety. Having said that, one of our interviewees (from a large oil company) readily admitted that if a nightmare scenario emerged, "All bets are off"; in such a case, social and political pressures might lead to rapid and discontinuous technological change with little regard for short-term costs. However, in the absence of such a threshold event, corporations will proceed on a path that they believe best balances social and environmental responsibility with economic and political reality.

Nevertheless, the Shell scenarios speak to significant technological and market-related change and uncertainty and so the likely defining importance of political, social, and economic investment factors in determining outcomes. One of these factors is the attitude of policy-makers and of society at large in supporting technological change and new political investments as alternatives to "business as usual." A second is the attitude of the funders of the commercialization of new technologies — the financial institutions and especially the venture capital community. Let us turn first to political and social attitudes.

Political and Social Attitudes

Clearly the enactment of the Kyoto Protocol and its future implementation have dominated political, public, and corporate debate on climate change. But even though there is now a framework in place for globally co-ordinated action on climate change, there remain many public policy options available to address the obligations created by Kyoto at a national level.[24]

Some parts of the Canadian business sector did not distinguish themselves in their immediate response to Prime Minister Jean Chrétien's announcement in September 2002 at the Johannesburg World Summit that Canada would indeed ratify the treaty. And perhaps it is kindest to draw a veil over the more lurid predictions of commentators and lobbyists that campaigned vociferously — against public and political sentiment — for the decision to be overturned. Happily, it is now well understood by enlightened political, business, and civil-society leaders in most parts of the world that climate change is real and that it merits a serious response by everyone, including business.

For example, when the U.K. Department of Trade and Industry released its energy white paper in February 2003, it included scenarios and business responses not unlike those described by Shell. The government envisaged a wide range of market-based initiatives to 2020, including enhanced energy efficiency by households and industry, voluntary agreements on vehicles, promotion of bio-fuels and other renewables, and support for the E.U. carbon-trading scheme. Taken together, these initiatives would achieve cuts of 15–25 million tonnes of carbon while providing significant support for renewable and distributed energy suppliers ($1 billion a year by 2010), fuel-cell research, and low-carbon innovation.

Despite the somewhat conservative public policy pronouncements emerging federally (if not at the state level) in the United States,[25] we believe it would not be unreasonable to expect Canada to behave more like the United Kingdom, the Netherlands, Germany, Japan, and many other countries in mobilizing a proactive, inclusive, business-friendly approach to climate change. Indeed, we are seeing such partnership-based strategies emerge even in the United States, if not at the federal level, then certainly from state governments, from civil-society organizations and think tanks, and from business itself (for example, from business and state coalitions committed to purchasing green power).[26] Based on recent documents issued by federal and provincial governments in Canada,[27] there does seem to be a serious intent to reduce Canada's net greenhouse gas emissions by up to 240 million tonnes by 2010.[28] And there are excellent examples of serious business engagement with climate-change solutions in recent years, for example in the fields of greenhouse gas emissions trading and other voluntary initiatives.[29] We will describe these in some detail later. For now, it is worth noting that in our research for this chapter we detected some unease in Canadian business circles with respect to exactly how seamless public policy would be in practice given the traditional federal and provincial rivalries.

Investment Climate

On the question of providing funding for market-relevant innovation, it is clear that the Canadian government sees support for corporate R&D and basic research as vital components of its approach to addressing climate change.[30] However, it is unusual for governments to risk tax dollars on taking R&D to market via new ventures. This is the gap filled by venture capital investors, a group whose attitudes to the risk and reward equation will be of fundamental importance to the emergence of new technologies aimed at mitigating climate change.

Joel Makower and colleagues at Clean Edge research have drawn attention to the likelihood that in the United States, "many early-stage companies will likely wither on the vine for want of consistent policies and sufficient capital. . . . This represents a lost opportunity."[31] Nevertheless, Makower and colleagues claim that solar photovoltaic production will increase globally from US$3.5 billion in 2002 to US$27.5 billion by 2012, that wind power will expand even more — from US$5.5 billion to US$49 billion in the same timeframe, and that hydrogen fuel cells (for all applications) will grow from US$500 million to US$12.5 billion. They note that in 2002, energy technologies represented 2.3 percent of total U.S. venture capital activity compared with 0.9 percent in 1998.[32] Of course, it is not just wealthy individuals and established venture capital players who are active in supporting the development of novel technologies. The practice of "corporate venturing" is now expanding rapidly to enable established energy and non-energy players to obtain a window on the technologies of the future through shared investments in venture funds — often with their potential competitors as well as their business partners. We will return to this phenomenon when we discuss issues of corporate re-invention and the avoidance of the "innovator's dilemma," a term that describes the tendency of large incumbent businesses to miss new applications of technology.[33]

But let us now take a tour of both today's and tomorrow's strategic options for business in "fueling the future." We will do this from the perspective of internally and externally driven initiatives, and so will describe four categories of business innovation in energy production and use. In doing this we are following the logic of Stuart Hart of Cornell University, one of the world's leading thinkers on sustainable business strategy, who has developed an influential model for generating "sustainable value" that considers these four dimensions.[34] Our four categories are

- Reducing Risks and Improving Profits: Driving Air Pollution Down and Energy Efficiency Up
- Creating Value for all Stakeholders in a Carbon-Constrained World
- Innovation for the Products of the Future: Incremental Change or Strategic Reinvention?
- Securing Growth in the Markets of the Future

We will start with the most obvious strategy, in which businesses save money and reduce their environmental liabilities today through simple internal process improvements and plain good housekeeping, thus driving air pollution down and energy efficiency up.

REDUCING RISKS AND IMPROVING PROFITS: DRIVING AIR POLLUTION
DOWN AND ENERGY EFFICIENCY UP

We must understand that we do have a daily impact on the environ-
ment. Maybe there is not enough evidence to convict, but there is enough
evidence to go to the grand jury of public opinion.
— Erroll B. Davis Jr., Chairman, President,
and CEO, Alliant Energy[35]

In our research for this chapter, many of our business interviewees
mentioned the effective delivery of energy efficiency and thus
reduced climate-change impacts by a range of international and
domestic companies. The most frequently mentioned companies
were BP, DuPont, Dofasco, Royal Dutch/Shell, and TransAlta.
However, the full list of citations included the following: oil
and gas companies BP, Shell (including Shell Canada), Syncrude,
and Suncor; heavy industry and metals companies Alcan, Alcoa, and
Dofasco; chemicals companies DuPont, Dow, BASF, and Nova;
manufacturers Interface Canada and Husky Injection Molding; and
a number of Canadian energy utilities including BC Hydro, Ontario
Power Generation, TransAlta, Manitoba Hydro, and Quebec
Hydro. The mining and forestry sectors in Canada were also cited
as very active in driving energy efficiency, with companies such as
Inco, Noranda/Falconbridge, Placer Dome, Nexfor, Abitibi, and
Domtar all mentioned by our interviewees.[36] Clearly something
significant is happening when so many businesspeople are able to so
readily identify leadership activity in so many diverse companies
and industries. The underlying logic here is quite simple.

Energy efficiency represents a "win-win-win" for the environment, human health, and the profitability of firms. Many companies have successfully reduced their energy consumption and
thus their greenhouse gas and smog-producing emissions. In many
cases the solutions are not especially "high-tech" or necessarily

based on new technological solutions, but rather smarter ways of doing things, employing the ingenuity and common sense of managers and employees. Examples include the introduction of less carbon-intensive power plants that use combined cycle gas turbine power, the capture of landfill gas emissions, and the application of industrial co-generation technologies. For example, co-generation, also called combined heat and power (CHP), is the simultaneous production of thermal and electrical or mechanical power from the same fuel in the same facility. It is often achieved through the capture and recycling of rejected heat that escapes from an existing electricity-generating process. This combination typically achieves energy efficiencies of up to 80 percent and is technologically straightforward and undemanding to finance and install.

The Canadian Industry Program for Energy Conservation (CIPEC) represents 5,000 companies organized in forty-three trade associations contributing to twenty-five task forces aimed at driving and learning from "best practice initiatives" in energy efficiency. Supported by Natural Resources Canada, but industry led, CIPEC produces an annual summary of *Success Stories* that demonstrates the opportunities for economic, environmental, and social gain through energy-efficiency initiatives.[37] Some of the examples cited[38] in 2002 included Ontario Power Generation's saving of 2.7 million tonnes of CO_2 through 300 separate energy efficiency projects, and Nexen's saving of 850,000 tonnes of CO_2 equivalent per annum through better management of methane gas emissions.

In Canada, as in many other countries, voluntary schemes employing reporting, target setting, and peer pressure have been introduced in order to provide a framework for companies to measure and reduce greenhouse gas emissions.[39] The Voluntary Challenge and Registry (VCR) program is a key part of the public policy landscape for optimizing the business response to climate change in Canada. More than 900 organizations are currently registered with VCR, covering a large proportion of "large industrial emitters." Since 1997, VCR, a multi-sector initiative supported by the private and public sectors, has encouraged effective reporting and target setting at the organizational level and has recognized those achieving significant progress through Leadership Awards backed personally by the federal Natural Resources and Environment ministers.[40]

One of the most effective techniques for reducing greenhouse gas emissions that has been employed by a number of leadership companies in Canada is "performance contracting." This technique can provide an effective means of implementing energy efficiency projects while avoiding upfront capital costs. The company, usually working with an external third-party energy services company or ESCo, simply finances its energy efficiency investments "off the balance sheet," thereby avoiding institutional barriers to capital spending. DuPont Canada, Ford Canada, and Inco have all successfully used this strategy to execute a number of energy efficiency projects. Following the achievement of aggressive corporate targets for energy efficiency worldwide during the 1990s, DuPont Canada is now targeting 6–8 percent total energy savings across the company based on energy efficiency investments made through its two plants at Maitland and Kingston. At Maitland, a $15 million investment resulted in savings of $2.5 million and 30,000 tonnes of CO_2 per annum.[41] Inco's approach to reducing energy consumption harnesses the ingenuity of all of its employees through a program called "Power Play." Since 1999, it has saved over $20 million in energy expenditures as a result of aggregate savings from hundreds of simple ideas suggested by its employees.

For some sectors, such as the coal-fired electricity producers, the challenge goes well beyond energy efficiency. It may be a question of life or death for their industry. The Canadian

Clean Power Coalition (CCPC) is a group of coal-fired generating companies that recognize the imperative of significant improvements in the technical efficiency of their plants and processes if they are to retain any hope of being part of a carbon-constrained energy sector in the future. Thus the CCPC plans to invest up to $1 billion over the next two years in the construction of two "clean coal" demonstration plants with even lower CO_2 emissions than those of combined cycle gas turbines.

Internationally, the Pew Center on Global Climate Change has published detailed case studies of six companies with comprehensive approaches to targeting greenhouse gas emission reductions.[42] The six companies are Asea Brown Boveri, Entergy, IBM, Toyota, United Technologies, and Shell (see box for details of the Royal Dutch/Shell case). In each case there were different reasons for adopting such comprehensive strategies. All believed that setting targets for reducing greenhouse gas emissions would improve competitiveness and enhance sales. Other reasons included a desire to prepare for future regulation, to contribute to the development of fair and efficient international and domestic policies, and to generate reputational gain. The companies recognized that there were risks associated with prompt action. For example, governments may fail to give credit for early action or they may set baselines that do not sufficiently reward early movers. Or they may not regulate at all, in which case companies will have incurred costs for no long-term benefit. We will return to this issue in our conclusion.

The pursuit of energy efficiency has resulted in significant gains for some international industry sectors. For example, according to the Organisation for Economic Co-operation and Development (OECD), between 1985 and 1996 the E.U. chemical industry realized an energy efficiency gain of 34 percent per unit of production. Between 1974 and 1998, the U.S. chemical industry achieved a 43-percent gain. And between 1971 and 1991, the steel industry in ten OECD countries

ROYAL DUTCH/SHELL AND SHELL CANADA: AGGRESSIVE TARGETS AND SYSTEM-ATIC FOLLOW-THROUGH

Despite the fact that Shell's greenhouse gas emissions were 111 million tonnes in 2000, making it one of the world's largest industrial emitters, the company is a success story in managing its own direct emissions. In 1998, the company set a goal of reducing its greenhouse gas emissions to 10 percent below its 1990 baseline. It achieved that goal in 2002. The reductions were largely a result of phasing out continuous venting and flaring of gas during oil production, but increased operational efficiency was also a factor.

There have also been a number of successful eco-efficiency pilots. One, in Fredericia in Denmark, took the second most efficient oil refinery in the Shell system and identified potential energy savings of 17 percent and cost savings of US$6–7 million. Shell's current target is to keep greenhouse gas emissions at 5 percent below 1990 levels despite its significant business growth. Shell's ability to reduce emissions and energy consumption can be attributed in part to the application of "carbon shadow pricing" to any new ventures. All projects that come to the Shell board for approval must include Shell's forecast future carbon costs in their base assumptions.

Another energy-conscious mechanism that Shell established is its internal Tradable Emissions Permit System (STEPS), where it has gained experience with emissions trading and found least-cost opportunities for emissions reduction. In tune with the priorities of its parent, Shell Canada has set emission reduction targets aligned with Canada's Kyoto target of 6 percent below the 1990 baseline by 2008. The company has been making ongoing investments in energy conservation measures since the mid-1990s. Its projects include installation of waste-heat recovery mechanisms, variable frequency drives, and hydraulic power recovery turbines at various Canadian facilities.

achieved a 20-percent improvement in energy efficiency.[43]

The successes of sectors and individual companies in reducing energy use and improving energy efficiency are in many cases impressive. However, as we noted earlier, because populations and economies continue to grow and industrial production keeps pace, gains in efficiency — if they continue — will have the effect only of moderating global demand for primary energy from three times to two times the current demand by 2050. Thus, we may conclude that energy efficiency gains based on technological improvements are important, but they provide an insufficient basis for addressing the immediate climate-change imperative for business — internationally and in Canada. Certainly a good deal more than incremental improvements in energy efficiency will be required if Canadian large industrial emitters are to deliver a reduction of 55 million tonnes of CO_2 by 2010 as envisaged by the federal government.[44]

Now we will turn to the next element of our analysis, which considers how external political, social, economic, and environmental forces affect business today and how business must respond if it is to continue to drive sustainable value in the immediate future.

CREATING VALUE FOR ALL STAKEHOLDERS IN A CARBON-CONSTRAINED WORLD

Throughout our journey we have made significant strides in understanding what sustainability means, and how we can apply it to providing energy solutions that meet or exceed the environmental, economic and social needs and expectations of our stakeholders. We have also become more aware of how much we still have to learn, and how far we still have to go. However, each step we take helps us to earn the trust and respect of stakeholders and supports our ability to operate responsibly, grow our business and generate shareholder value.

— Rick George, President and CEO, Suncor Energy

Every country that has ratified the Kyoto Protocol is seeking the most economically efficient and socially acceptable way to meet its greenhouse gas emission reduction targets by 2008–2012. As noted above, technologically driven improvements in energy efficiency in key industries will continue to be important in the short term. But it is likely that significant external drivers will be necessary to secure hoped-for reductions in greenhouse gas emissions from business and industry. Two of the most important drivers for business are (a) public policy related, and (b) investor related. These two drivers are converging, as we will demonstrate below.

Public Policy Drivers

Governments around the world are taking a variety of complementary approaches to reducing emission intensities in their countries, including demand-side management,[45] industry covenants, fiscal incentives, carbon taxation, and emissions trading. The Climate Change Plan for Canada proposed five key elements for achieving the national goal "for Canadians to become the most sophisticated and efficient consumers and producers of energy in the world and leaders in the development of new, cleaner technologies."[46] These elements are:

- Emissions reduction targets for large industrial emitters established through covenants
- A partnership fund to share the burden of implementation between all actors in Canada
- Strategic infrastructure investments (e.g., urban transit projects)
- A co-ordinated innovation strategy
- Targeted measures including information, incentives, regulations, and tax measures.[47]

Each of these elements sends important signals to business. Several offer potential incentives for companies willing to innovate and take advantage of climate-friendly opportunities. There are particular opportunities and challenges for the transportation and building sectors, and it is clear that the federal government will be investing significant resources in promoting energy-efficient public and private transportation and energy-efficient commercial and residential property.[48] But perhaps the most immediate and high-stakes issue is the question of targets for large industrial emitters linked to industry covenants and carbon emission caps and emissions trading for key sectors.

Market-based approaches to encouraging industrial energy efficiency are emerging as the most popular mechanisms for reducing greenhouse gas emissions within the private sector. Emissions trading enables participants to find the most cost-effective and flexible means of reducing emissions by buying and selling credits.[49] From a business standpoint, flexibility is desirable because it allows for cost-effective means to lessen the burden of compliance with targets and obligations. The Canadian government proposes to establish covenants, backed by regulatory measures, with nine industry sectors selected on the basis of emissions intensity. The sectors are thermal electricity generation, oil and gas, mining, pulp and paper, chemicals, iron and steel, smelting and refining, cement and lime, and glass and glass containers.

The Dutch government initiated a covenant-based system for constraining carbon emissions by industry in 1999. Originally negotiated with six industry associations, the Energy Efficiency Benchmarking Covenant requires companies to aim for top-ten percentile performance worldwide for industrial processes. In support of this goal, companies needed to benchmark and then report on all aspects of energy management. By 2002, the scheme was judged a success by the Dutch government, with ninety-seven industrial companies and six power-generating companies engaged through 232 sites and 528 different processes.

The sectors involved included iron and steel, non-ferrous metals, brewing, cement, chemicals, glass, paper mills, sugar, and electronics. Together these companies represented 84 percent of the possible participants; based on the submitted energy efficiency plans, they were scheduled to deliver savings of 4.6 million tonnes of CO_2 by 2012.[50]

Similar to the successful experience in the Netherlands, the process by which sector targets will be negotiated in Canada will be based on an analysis of current CO_2 intensity per unit of product, comparing that with "best available technology" intensity, applying a target reduction (a minimum 15 percent is suggested), and allocating an allowance for the sector. The gap between the target and the allowance (the shortfall) then becomes the basis of the market in carbon.

For example, if a sector such as cement is 20 percent less efficient per tonne of cement produced in Canada than in Europe, and if the sector produces 1 million tonnes of CO_2 per annum, the government might provide it with an "allowance" of 800,000 tonnes per annum to be divided up among the companies in that sector. In this case the government knows that it is technically possible for the 200,000 tonnes to be saved, and that these savings fit with the overall strategy of a minimum improvement of 15 percent per sector. Companies that represent 10 percent of the cement sector would then have an allowance of 80,000 tonnes.

Imagine two companies, each representing 10 percent of the industry. Company A invests in energy efficiency measures and becomes so efficient that it produces only 70,000 tonnes of CO_2 per annum; it then has a credit of 10,000 tonnes that it can trade on the open market. Company B, of equal size, does not invest and still produces 100,000 tonnes per annum; it then has the choice either to make the same investments as Company A or to buy rights to emit 20,000 tonnes more than it is otherwise allowed. In this way companies that stay within their allowance or buy carbon credits from others in a domestic or international market will be exempt from any penalties imposed by the government for non-compliance. Companies that do not require all of their credited reductions in emissions may sell them and make money.

The U.K. has operated an emissions-trading program since January 1, 2002, with 5,000 entities now engaged in the system. The E.U. is preparing to open the first international emissions-trading system by 2005, at which time 10,000 steelworks, power generators, oil and gas refineries, paper mills, glass factories, and cement installations will be affected.

In March 2003, the Chicago Climate Exchange (supported by BP and Suncor) opened the first North American greenhouse gas emissions–trading program. Large firms such as Shell and BP have also instituted intra-firm carbon markets in addition to factoring the cost of carbon into capital expenditure decisions on a project life-cycle basis. This allows their capital planners to compare the likely cost of carbon on the open market with the costs of energy efficiency investments and to choose whichever is the most economically attractive option (a practice called "shadow pricing"). According to the World Bank, even before a functional international carbon emissions-trading system is in place, a growing number of such transactions is occurring. Between 1996 and 2001, at least 55 million tonnes of greenhouse gas emissions were traded in more than sixty-five trades.[51] Many of these trades were driven by companies wishing to hedge against a rise in prices of carbon in years to come (current prices for a tonne of CO_2 equivalent rarely exceed US$5).

In decades to come, the price of carbon will be determined by the market, which in turn will be driven by the effectiveness of the implementation of the Kyoto Protocol and its manifestation in different jurisdictions. Few expect the price of carbon to exceed $15 per tonne in the short to medium term, and thus the Canadian federal government is on reasonably safe ground in providing a guarantee to business that any costs above this figure will be met by the government. It was this guarantee, along with Suncor's announcement that implementation of the Kyoto Protocol would not cause significant problems for their business, that effectively deflated business opposition to Canada's policy on climate change.[52]

Investor-Related Drivers

Just as influential for business as any public policy drivers are the constant pressures from investors for return on investment and avoidance of unnecessary costs of doing business. Thus it is interesting to observe a convergence of investor, government, and societal interests on the question of climate change. One notable recent development in this regard is the emergence of the Carbon Disclosure Project (see box); another is the emergence of shareholder resolutions on the extent to which companies are exposed to risk as a result of the carbon intensity of their operations.[53]

Increasing numbers of investors and financial institutions are starting to measure companies' exposure to climate change risks. A report published by the World Resources Institute (WRI) in 2002 was one of the earliest in-depth analyses of the comparative exposure of a range of oil companies to climate change policy and environmental/community restrictions on access to reserves. *Changing Oil: Emerging Environmental Risks and Shareholder Value in the Oil and Gas Industry* estimated ranges of possible outcomes for sixteen oil companies. Among the least carbon-intensive — and therefore least exposed to climate change risk — were Burlington Resources and Sunoco. Among the most exposed were Occidental, Repsol YPF, and Unocal. Potential negative impacts on shareholder value ranged from 0 to –11 percent, with most companies' median estimates being in the range of –2 to –7 percent.[54] This means that individual investors buying oil stocks might be well advised to avoid the most carbon-intensive and therefore risk-exposed companies if they wanted to maximize their investment gains from the sector. Eventually, when a large number of fund managers become aware of these facts, risk-exposed companies will find it harder to raise capital and the better performers will gain competitive advantage.

Also in 2002, the Coalition for Environmentally Responsible Economies (CERES) published a report by Innovest Strategic Value Advisors entitled *Value at Risk: Climate Change and the Future of Governance*, which drew attention to the fiduciary duty and governance implications of carbon risk management for corporate directors and institutional investors.[55] The authors drew attention to the nature and size of risks — and opportunities — arising in sectors such as transportation, water and waste, petroleum, gas and pipelines, forestry, pulp and paper, basic industries (steel, chemicals, and mining), tourism, building construction and real estate, manufacturing, agriculture and food, insurance, and electric utilities. In the latter case, which was given special attention, the authors estimated that the discounted future costs of stabilizing greenhouse gas emissions at 1998 levels for thirty-one U.S. utilities ranged from US$1.2 billion for the most exposed firm (American Electric Power) to US$51 million for the least exposed (Public Service Enterprise Group) based on a cost of US$20 per tonne of carbon. Again, the relevance for future investors in American Electric should be obvious; here is a company that will suffer disproportionately as public policy starts to take effect and the company's carbon emissions become an increasingly expensive liability.

The Carbon Disclosure Project (CDP) is a logical development of the WRI and CERES reports in that it provides a mechanism for the investment community to obtain objective evidence of the risks associated with carbon intensity in the energy industry. As a mechanism predicated on transparency and accountability to shareholders, the CDP provides an excellent example of the increasing convergence of public policy and environmental and investor interests in the context of climate change (see box).

THE CARBON DISCLOSURE PROJECT: A VIRTUOUS CYCLE OF TRANSPARENCY, ACCOUNTABILITY, AND THE EFFECTIVE MANAGEMENT OF RISKS FOR INVESTORS AND THE ENVIRONMENT

The Carbon Disclosure Project (CDP) is a co-ordinating secretariat for institutional investor collaboration on climate change. The driver for this project was a realization by the investment community that the potential carbon liabilities within a single emissions-intensive firm could represent as much as 40 percent of its entire market capitalization under certain scenarios. On the other hand, companies that manage this risk effectively will gain competitive advantage.

In May 2002, a letter was sent to the chairs of the boards of the 500 largest companies in the world (by market capitalization) on behalf of thirty-five institutional investors who jointly represented assets in excess of US$4.5 trillion. In this letter, the companies were asked to identify the business implications of their exposure to climate-related risks and to explain what they were doing to address these risks. Two hundred and twenty-one companies completed the questionnaire and fifty-eight others responded in some manner. The key findings of the CDP survey indicated that over 50 percent of FTSE 500 companies recognized climate change as a serious issue and were developing (if not implementing) strategies for reducing greenhouse gas emissions.

The companies recognized financial risks posed by climate change from (a) the direct impact of climate change itself, and (b) exposure to the costs of greenhouse gas emissions from

---

regulatory regimes trying to mitigate climate change. The impact of climate change will vary geographically, between sectors, and between companies within those sectors. But the CDP highlights the fact that the financial risks and consequences of climate change are real and will likely intensify, and that winners and losers will emerge. Thus investors are demanding greater transparency and disclosure so they can manage their own risk. We understand that the CDP will be repeated in 2003/2004.

Value for Stakeholders

It remains to be seen whether signatories to the Kyoto Protocol will deliver on their 2008–2012 commitments. If they do, it will be due in no small measure to the ability of business organizations to accelerate their energy efficiency efforts, encouraged by governments and investors responding to global societal priorities. These efforts are *de facto* incremental — improvements across different industry sectors averaging 15 percent in the case of Canada. In achieving these gains, signatory countries may suffer minor impacts on GDP growth (in the case of Canada, perhaps between –0.4 and –1.6 percent to 2012). But the opinion of the E.U. and Canada is that these impacts will be more than compensated for by improvements in economic efficiency and human well-being. Thus, we would argue that Kyoto, and its ushering in of a more carbon-constrained global economy, does represent "stakeholder value" in the sense that people and businesses will still make money, GDPs will be largely unaffected, and yet there will be social and environmental gains associated with implementation.

But as we saw earlier in our discussion of energy scenarios, Kyoto is only the starting point. For the world to transition to a truly sustainable energy future, significant leaps forward will also be needed in terms of products, services, and consumer behaviour. A good deal of this change will depend on the ingenuity of business to research and develop the required products and services; meanwhile, governments will need to provide the right incentives, and consumers will need to send the right market signals.

We will now turn to the medium- to long-term future and explore possibilities for the emergence of radically improved products and services with significantly less climate change impact and an ability to displace existing products and processes altogether.

INNOVATION FOR THE PRODUCTS OF THE FUTURE: INCREMENTAL CHANGE OR STRATEGIC REINVENTION?

In the medium term, we see solar energy as both a profitable growth business and an important part of the effort to reduce our greenhouse gas emissions. Within our growing Renewable and Alternatives business, we believe that by continuing to focus on reducing costs and creating new commercial products, both solar and wind power will help us meet the world's demand for sustainable and affordable energy long into the future.

— John Mogford, Group Vice-President,
Renewables and Alternatives, BP

There is a growing belief in some opinion-forming circles that business and industry are capable of the sort of technological and industrial process innovation required to address and arrest climate change. Techno-optimists like Amory Lovins believe that climate change can be addressed through harnessing industrial ingenuity rather than the imposition of legislation or social control. Much of their rhetoric is about "opportunities" rather than "threats" or moral obligations, and is thus is not at all palatable to many in the environmental community.[56]

Earlier in this chapter, we described some of the challenges that small and large businesses face in mobilizing capital to support innovative products that will transcend existing energy-intensive or inefficient products. In his book *The Innovator's Dilemma*, Clayton Christensen made the general point that larger, incumbent firms often find it especially difficult to develop products that will undermine their existing offerings. For example, researchers in an automotive firm might understandably have an inbuilt bias towards the internal combustion engine, which has served the corporation so well for a century. Thus, it is argued, any attempt within that firm to promote innovation with fuel cells will be strangled by all kinds of internal institutional forces. This phenomenon explains — at least in part — why increasing numbers of large firms are deciding to invest in new, more sustainable technologies at arm's length through a process described as "corporate venturing." Indeed, that is precisely why Ford, DaimlerChrysler, and GM have invested in Canadian hydrogen fuel-cell companies like Ballard and Hydrogenics. Nevertheless, there are businesses — large incumbents and smaller start-ups and SMEs — that demonstrate that a re-invention of products and product portfolios consistent with a sustainability strategy may be achieved internally, delivering value for both investors and society. One company with a very clear product-led sustainability strategy is DuPont.

DuPont employs a number of approaches to work towards a more sustainable product portfolio.[57] Using an explicit frame of "sustainable growth," the company is leveraging years of expertise in electrochemical technology, polymers, and coatings in order to manufacture components for fuel cells — a technology that might itself revolutionize energy and power markets in the future.[58] DuPont is working with several industry partners to reduce costs, develop competencies, and increase the availability of fuel-cell technology. DuPont is also committed to life sciences and is employing biotechnology to develop novel plastics, paints, and fibres, thus reducing petroleum-based feedstocks. Meanwhile, DuPont has steadily been divesting carbon-intensive industries such as oil and nylon manufacture where the nature of being in a commodity business leads to unpredictable returns as well as significant greenhouse gas liabilities.

DuPont, like many chemical companies, is highly dependent on the future of the automotive industry. And this industry is also being driven by public policy and investor-related pressures to radically reduce its carbon intensity — both in manufacturing and in terms of the products it develops. For example, the Canadian government expects to negotiate a 25-percent improvement in new-vehicle fuel efficiency by 2010, which automotive manufacturers would achieve through existing and expected new technologies. Many commentators believe that high-efficiency diesel engines and hybrid engines may provide many of the answers here — at least in the short to medium term. The development of bio-fuels in order to supplement gasoline or diesel may also have significant merit and thus the federal government is hoping to support both.

Many combinations of hybrid vehicles are possible, including gasoline, natural gas, or hydrogen engines with an electric motor.[59] The majority of current designs are based on compact or subcompact formats, which have been pioneered by Japanese manufacturers (see box). However, some medium-duty hybrid vehicles are currently available, and in May 2003, FedEx announced plans to integrate twenty hybrid-electric diesel delivery trucks by early 2004. FedEx expects to purchase hybrids on the company's regular purchasing schedule for their medium-duty delivery trucks. They have the potential to replace the company's 30,000 delivery trucks over the next ten years.[60]

INNOVATION FOR THE PRODUCTS OF THE FUTURE: TOYOTA AND HONDA LEAD THE WAY ON HYBRID AUTOMOBILES

Significant gains can be made in automobile fuel efficiency, and hybrids are currently the most promising technology on offer for mass commercialization. Hybrid vehicles draw power from two different energy sources, usually a gas or diesel engine with an electric motor. Presently Toyota and Honda are the leading car manufacturers with hybrid offerings in North America. Honda was the first to market with the Insight in 1999, which is still the most fuel-efficient vehicle on the road today according to NRCan's EnerGuide awards, achieving 3.9 L/100 km in the city and 3.2 L/100 km on the highway. Honda also offers the Civic in a hybrid model. The Civic is more comfortable and roomier than the Insight, but compromises on efficiency, getting 4.9 L/100 km in the city and 4.6 L/100 km on the highway. Toyota's Prius gets slightly better mileage than the Civic and has been the best-selling hybrid in Canada since its release in 2000.

North American automakers have announced plans to launch hybrid models in the near future. They have chosen a different strategy by introducing pickups and SUV hybrids to the North American market. GM promises hybrid Silverado and Sierra pickup trucks by the end of 2003. Ford will put a hybrid option of the Escape on the market in summer 2004. Plans for other domestic hybrid automobiles will be phased in over the coming years.

Another part of the vehicle fuel-efficiency innovation story is the reduction in weight of automotive bodies without the loss of structural integrity. Here it is interesting to note that a Canadian steel company has been at the forefront of innovation for product sustainability.

Dofasco is one of Canada's largest steel producers, and one of the most successful steel companies in North America. Recognizing that in order to stay competitive in tomorrow's markets it would have to change the way it does business, the company adopted an innovation strategy in the early 1990s to differentiate itself in a competitive and global marketplace. It has evolved from a commodity producer to a steel service provider, focusing on value-added products like the Ultra Light Steel Auto Body (ULSAB). This project was initiated by Dofasco and US Steel, which were soon joined by thirty-three additional steel-making partners around the world. The ULSAB project resulted in a number of product developments that contribute to light-weight, more fuel-efficient vehicles.

In the energy supply industry, it is questionable whether we should describe all renewable and alternative sources of energy as innovations or "products of the future," because in many cases the technologies already exist. It is more an issue of economics as to if, how, and where installations may occur in the future. The major exception to this is the potential for future breakthroughs in solar energy using thin-film technology based on inorganic, organic, and polymer-based materials.[61] However, there is no doubt that wind and solar energy will grow in importance in both developed and developing countries even with existing technologies. In some circumstances, they may even represent a "disruptive technology" — replacing the need for conventional energy technologies altogether — especially if future technological innovations can be married with significant reductions in manufacturing costs.

The International Energy Agency predicts that by 2030 renewables (excluding hydro-electric energy) will contribute 4 percent of world energy needs — double the percentage in 2000 — and the World Energy Council predicts that the global market for renewable energy is likely to be in the range of

US$234 to US$625 billion by 2010 and perhaps as high as US$1.9 trillion by 2020.[62]

One of the most promising technologies for the longer term is solar energy. Photovoltaic (PV or solar cell) systems convert sunlight directly into electricity. Energy derived from solar power is renewable and the source is free. The capital cost of solar power for large-scale supply has historically been deemed prohibitive. However, direct manufacturing costs (in 2002 dollars) dropped from US$5.47 per peak watt in 1992 to US$2.42 per peak watt in 2002.[63] The life-cycle cost of solar energy generally ranges from US$0.20 to US$1.00 per kilowatt-hour.[64] With rising production levels, developments in efficiency, and breakthroughs in energy storage, solar power will increasingly become a viable energy source in off-grid applications. However, for conventional applications solar is probably still ten to twenty times too expensive compared with grid electricity, and thus significantly more ingenuity and technological innovation will be required before it can compete directly with fossil fuels for industrial and domestic use in industrialized countries.

Alongside Sharp, Kyocera, and Sanyo of Japan, BP Solar is one of the world's largest solar companies, with global market share of nearly 20 percent and ambitions to reach sales of $1 billion per annum by 2007. BP manufactures crystalline silicon solar cells and thin-film photovoltaics in plants around the world. The race is on in the PV world to increase efficiency of solar cells and decrease manufacturing costs. Innovative technologies that integrate PV into building materials such as glass, shingles, and siding may add new value to existing products as well as overcoming current architectural and aesthetic barriers. Further advances in lighting technology, such as white-light LED lamps, may facilitate wide-scale application of solar power in low-voltage systems in remote applications and developing countries.

Wind is another form of power currently attracting the

interest of major industrial players. In May 2002, GE Power Systems acquired Enron's wind assets. The addition of GE to the wind energy sector marks a coming of age for the industry. GE cites three main reasons for entering the industry: the cost of electricity generated from wind power dropped to the point where it was competitive with other sources; the business could benefit from technology from other GE businesses; and GE customers were increasingly interested in renewable energy sources.[65] GE Wind Energy now designs and manufactures wind turbines with rated outputs between 900 kilowatts (kW) and 3.6 megawatts (MW), sufficient to power several hundred homes. The 3.6 MW turbine is the first wind turbine over 3 MW designed specifically for the offshore generation market. It is currently being field tested off the coast of Spain and is expected to be commercially available in 2004. GE Wind Energy has received more that US$2 billion in orders and commitments since entering the industry in May 2002. The company estimates growth projections at around 20 percent annually. This will be helped in some measure by public policy–related targets. For example, the European Union is targeting a 12-percent renewable energy provision by 2010, and Denmark is aiming for 50 percent by 2030 — most of which is expected to come from offshore wind installations.

In all of these cases, large companies have taken their sustainability frame sufficiently seriously to develop products and technologies that at the very least may begin to erode sales of their traditional product offerings. These are typically large companies with a vision that extends beyond the short term. They also include investment banks like Merryll Lynch and Morgan Stanley that recognize there is money to be made.

In Canada, companies like Suncor, Shell Canada, TransAlta, Enbridge, and Manitoba Hydro are all making significant commitments to renewable energy development, despite the unfavourable electricity-price environment in most provinces. And a number of very promising renewable and alternative energy start-up companies in Canada are successfully attracting venture capital and even municipal investors.[66]

While Canada, with its abundance of cheap hydro-electricity and fossil fuel resources, is lagging behind many parts of the world in terms of installed renewable and alternative energy capacity, there may yet be hope for the flourishing of a more vibrant renewable energy sector in the future, especially if provincial policy-makers back aggressive targets as proposed by the Select Committee on Alternative Fuel Sources of the Legislative Assembly of Ontario.[67]

We now turn to the question of applying existing and future technologies to future markets in both the developed and the developing world.

SECURING GROWTH IN THE MARKETS OF THE FUTURE

I think there are few more challenging and worthwhile jobs in the world today than meeting the energy needs of a developing world in a sustainable way; few more stimulating than using technology and management innovation to solve fundamental problems — like tackling climate change — where creativity is embraced and applied.

— Sir Mark Moody-Stuart, Chairman,
Royal Dutch/Shell Group[68]

It is evident from much of what we have already described that the fundamental challenge of energy sustainability in the twenty-first century will be the degree to which the world's burgeoning population in developing countries will be able to leapfrog the inefficient, carbon-intensive technologies that characterized the industrialization of European and North American economies.

One of Shell's energy scenarios described earlier —

"Dynamics as Usual" — is characterized by two separate waves of growth in renewable energy, with the result that by 2050 renewable and alternative sources comprise perhaps one-third of primary energy needs. The second wave of renewable energy growth is led by breakthroughs in materials and solar technologies, so that in many applications they are "embedded." In this scenario, no spectacular growth of the hydrogen economy is envisioned and fossil fuels continue to meet the majority of primary energy needs.

In contrast, the second scenario — "Spirit of the Coming Age" — is characterized by breakthroughs both in fuel-cell technologies and distribution systems for hydrogen — the fuel of choice. From about 2015 there is a convergence on hydrogen fuel cells for both stationary and automotive applications, and from about 2020 there begins a significant leapfrog in energy technologies in countries like India and China, again based on a growing hydrogen economy. As in the first scenario, fossil fuels still provide the bulk of energy supply.

The radically different outcomes for renewables and fuel-cell technologies present energy-related businesses with a dilemma. Based on our many interviews with leading business-people, it is clear that no one is taking a definitive stance on those technologies that will "win" the race for energy markets of the future. We received comments to the effect that technologies like wind and solar "were not necessarily scaleable"; also, we got a sense that people were somewhat less convinced of the medium-term prospects for fuel cells than perhaps they were one or two years ago. Consequently, those companies at the forefront of thinking in this arena, including Shell and BP internationally and Suncor in Canada, are taking measured but bold steps in several directions — often through partnerships with venture funds, supply chain allies, and even competitors. For example, BP is partnering with DaimlerChrysler on a ten-city European Bus Project to explore the possibilities for hydrogen, with the

European Integrated Hydrogen Project on safe handling of hydrogen, and with the California Fuel Cell Partnership on commercialization of fuel-cell automobiles. Meanwhile, BP is hedging its bets by also investing seriously in energy efficiency, carbon sequestration, renewables, and natural gas.[69]

So when we consider markets of the future, we must also factor in the uncertainty of what might emerge, or might not emerge, as front-running technologies. We know that conventional renewables (including hydro-electricity and bio-fuels) have an increasing role to play,[70] and we know that under any scenario natural gas will have a bridging role in primary energy provision for the next few decades. We may also ponder the options for coal and oil and even the possibility of a renaissance for nuclear energy.

We considered renewables in the last section as potentially disruptive technologies that may or may not contribute to radical change in energy markets of the future — in both developed and developing countries.[71] Companies like BP and Shell are actively exploring what wide-scale off-grid renewable energy provision looks like in developing countries. BP has undertaken a solar-based lighting project in 50,000 homes in remote parts of Indonesia, and Shell, together with electric utility Eskom, has done a solar energy demonstration project involving thousands of homes in the Eastern Cape of South Africa.[72] In this section, we will keep these markets of the future in mind, but we will focus more specifically on prospects for the significant dislocation that would occur in all energy markets if the so-called hydrogen economy were to become reality.

Prospects for the Hydrogen Economy: The Ultimate Market Shift

In an ideal world, to be fully sustainable, hydrogen for fuel cells would be produced by renewable sources such as wind, solar, small-scale hydro, geothermal, tidal, biomass, or waves. During peak periods the energy produced from renewable sources would go directly to meet electrical demand, but at off-peak times the energy would be converted to hydrogen through electrolysis and stored for fuel-cell-generated heat and power, as well as for fuel-cell-powered transportation. Also, in an ideal world, hydrogen would be distributed safely in metal or composite tanks, or as metal or other hydrides, in nanotubes or in fibres.[73] If these conditions can be met, and if fuel-cell manufacturing costs can be constrained, then the prospects for the hydrogen economy meeting a large proportion of stationary and locomotive energy needs in the long term, for the markets of the future, are bright indeed.[74] Should this occur, the natural gas industry might still have a future, but the oil industry as we know it today would be effectively redundant within 50–100 years.

There are numerous reasons why fuel-cell technology and the possible advent of the hydrogen economy are attractive. To their obvious environmental benefits (reduction of urban air pollution, reduction of carbon emissions, reduction of rates of depletion of fossil fuels) may be added security benefits, flexibility and reliability benefits, and third world development benefits. This explains why the technology now receives active political and governmental support, notably in the United States, Japan, and the European Union, and why annual R&D investments and government support is running at about $1.5 billion per annum. U.S. President George W. Bush announced a US$1.2 billion investment in fuel-cell R&D in January 2003, specifically citing long-term energy security objectives. The European Union envisages fuel-cell investments of up to US$2.5 billion under its 2002–2006 Sixth Framework Program, with a target of 5 percent of road transportation to be fuel-cell powered by 2020. However, there are significant technological and infrastructural challenges to overcome if the hydrogen economy is to become a reality. There are simply too many large investments already sunk in the oil business to write off: oil wells, offshore platforms, tankers, refineries, and pipelines; not to mention downstream infrastructure: gas stations, automotive and engine plants, etc. Almost certainly, governments will have to be involved in helping create the conditions for new infrastructural developments, in both developed and developing countries.[75]

Nevertheless, it is worth noting that some of the leading R&D-based businesses in the hydrogen fuel-cell business are based in Canada. Two of the largest are Ballard, based in British Columbia, and Hydrogenics, based in Ontario. In addition, one of the leading venture capital firms in fuel-cell technologies is Chrysalix, which is based in Vancouver. And one of the world's leading suppliers of methanol — a strong candidate as a hydrogen "carrier" in early fuel-cell applications — is Methanex, also based in Vancouver. Meanwhile, Ontario Power Generation is planning to install the world's largest stationary combined fuel-cell heat and power unit on the campus of the University of Toronto (Mississauga) in the fall of 2003 with support from Siemens, Industry Canada, Natural Resources Canada, and the U.S. Department of Energy. It will be interesting to follow the fortunes of these pioneers in years to come — especially to see whether their initiatives and ingenuity can retain their independence, whether they can help create clusters of Canadian R&D-based businesses ushering in and feeding off the hydrogen economy, or whether their ideas and products are taken over by their U.S. customers and backers (see box).

FUEL CELLS: A UNIQUE OPPORTUNITY FOR TWO CANADIAN FIRMS TO LEAD THE WAY TO THE HYDROGEN ECONOMY

The most promising markets for fuel cells are in distributed power generation, locomotion, and niche applications. There are currently seventeen companies across Canada whose primary focus or goal is fuel-cell production and/or system integration. Canada is recognized as having world-class fuel-cell development companies like Ballard and Hydrogenics, both of which use proton exchange membrane (PEM) fuel-cell technology but which employ different strategies. Ballard Power, probably the best-known fuel-cell company in the world, was started in 1979. It has some high-leverage owners in Ford (20 percent) and DaimlerChrysler AG (24.2 percent), and, not surprisingly, Ballard's strategy centres on PEM fuel cells for the transportation industry. At Ballard's annual meeting in Toronto in May 2003, it showcased three Ballard fuel-cell-powered Ford Focus prototypes and invited shareholders to take a test drive. Ballard believes that it will take at least ten years before fuel-cell cars will be put into mass production. Hydrogenics also sells fuel-cell testing equipment to developers worldwide as a revenue generator, while simultaneously developing fuel-cell stacks, mostly for distributed generation applications and niche markets. Hydrogenics is also partly owned by GM, which holds approximately 20 percent.

It is perhaps a disappointment that there are no independent Canadian vehicle manufacturers taking a visible lead on designing the hydrogen fuel-cell-powered vehicles of the future. This would represent a major potential manufacturing and export opportunity for Canada. Instead, the major U.S., Japanese, and European manufacturers will likely become the leading customers for Canadian fuel-cell technology for automobiles. In theory, both Magna (an automotive manufacturer) and Bombardier (a public transit and airplane manufacturer) could leverage fuel-cell technologies in their own product designs for the future. It remains to be seen whether they will choose to lead or follow in this regard. Interestingly, Industry Canada has made innovation — including the promotion of fuel-cell technology — a key part of its 2003–2006 sustainable development strategy. It is to be hoped that this initiative will help support and stimulate strategic responses from Canadian firms well into the future. Clearly, the prize is there to be won: the estimated sales value of hydrogen-economy products supplied by Canadian companies for 2003 is only US$162 million, over 85 percent of which are exports. But by 2010 the global market may be as high as US$46 billion.[76]

Carbon Trading

In this tour of business opportunities in fueling the future, we would be remiss in omitting one very important market of the future for businesses seeking to operate competitively in a carbon-constrained world, and that is the market for carbon itself.

As a public policy tool, emissions trading is not without its critics, for example environmentalists who see the danger of large carbon emitters buying cheap credits from Russia and elsewhere while continuing with business as usual. However, a number of the businesses we spoke to in researching this chapter are actively exploring what the emerging market for carbon credits (as a commodity) might mean for them, and their level of enthusiasm is high.

Earlier we described how corporate energy efficiency strategies link to compliance with covenanted allowances and therefore the ability to trade surpluses or buy credits. What seems clear is that despite the challenges of allocating allowances, harmonizing trading, verifying claimed credits, minimizing transaction costs, and so on, there are powerful economic arguments for carbon trading being a highly efficient mechanism for allocating costs of compliance with the Kyoto Protocol.

The Pew Center on Global Climate Change has provided support for the argument that properly designed international emission-trading systems may significantly reduce the overall costs of carbon mitigation. In this case there would be benefits for both sellers and purchasers of carbon credits, provided that both international harmonization and flexibility are built in by allowing trades across different greenhouse gases, sources, and over time.[77] The Pew Center has also provided case studies of trading schemes that are under development at the international, national, and state levels and within corporations, and of actual trades and why they occurred. Two of these involved Canadian companies: TransAlta and Ontario Power Generation.[78] The authors estimate that sixty-five public trades of more than 1,000 tonnes of CO_2 equivalent occurred between 1996 and 2002, typically costing well under $3 per tonne.[79]

There is little doubt that trading carbon will become very big business for many companies and indeed national economies in years to come. Those now participating in prototypical trades are taking a risk, gambling that their trades made and "banked" at current relatively low prices per tonne of CO_2 will be recognized by trading systems in due course. However, they are also hedging against the likelihood that prices of carbon may be much higher in the future. For the short term, it is an interesting use of investors' money and it is to be hoped that companies that indulge in piloting such innovative solutions to future constraints on carbon emissions will not be punished either by politicians or by shareholders.

CONCLUSION: THE FUTURE BECKONS

In researching this chapter, we interviewed thirty Canadian and international businesspeople with in-depth knowledge of business and energy issues. Most respondents believed that significant progress is being made on energy efficiency in a number of sectors — driven largely by cost factors. As one senior energy industry player told us, "Altruism on its own is not economically viable." Many interviewees stressed that it is the heavy energy users — mining, metals, petrochemicals, pulp and paper, chemicals, and downstream oil and gas — who have tended to make most progress in recent years. In contrast, many people cited a range of sectors in which progress has been slower, such as thermal electricity, upstream oil and gas, automotive, residential and light industry buildings, conventional manufacturing, information and communications technologies, and pharmaceuticals.

When it came to looking forward twenty years, respondents were split quite markedly on which industry sectors would be more or less dependent on fossil fuels. For example, about one-third of our interviewees said that the automotive and transportation sectors would contribute most to a reduced dependence on fossil fuels over the next two decades. But an equal number saw automotive and transportation remaining dependent on fossil fuels for the foreseeable future. One interviewee advanced the interesting idea that some automotive companies "see hydrogen fuel cells as a threat" and that therefore they would block their introduction, locking themselves and the rest of the world into hydrocarbon dependence for competitive reasons. The oil and gas sector was also named in both categories (less dependent *and* still dependent) by several interviewees. Electricity generation — especially in North America and developing countries — was believed by many to be a sector highly wedded to carbon dependence because of the abundance of cheap coal in so many countries. Aviation, heavy manufacturing, and petrochemicals were all cited as sectors unlikely to reduce their carbon intensity significantly in the next two decades. Our interviewees frequently mentioned the residential housing and domestic appliance sectors as important components in reducing dependence on fossil fuels, in these cases actively supported by the purchasing behaviour of individuals.

What emerges from this picture is a sense of uncertainty and ambiguity for most of those we interviewed. There is certainly no consensus on what the future might look like, or whether some industry sectors will be more or less dependent on fossil fuels in coming decades. Virtually everything is contingent on external factors and technological progress.

There are significant concerns about how the external factors will play out. For example, in terms of societal attitudes, it is entirely unclear to businesspeople how individual citizens will play their part in sending positive market signals to companies. And in terms of political decision-making, we noted a number of important caveats in the context of energy efficiency and the development of new markets for low or renewable energy products and services, internationally and domestically.

These sources of external uncertainty are worth summarizing:

- The risk that governments internationally may fail to give credit for early action;
- The risk that those participating in early trades will be punished;
- The risk that governments internationally will not help create the conditions for effective infrastructural developments, in both developed and developing countries;
- The risk that positive market signals will not materialize — through public and private sector purchasing or through consumer markets;
- The risk that cheap energy prices in Canada will stifle innovation in alternative energy technologies; and
- The risk that lack of federal–provincial co-ordination will further hamper prompt action.

Taken together these sources of risk may effectively negate the possibility of establishing Canadian leadership through effective R&D, venture financing, and commercialization of new technologies, with the result that current Canadian leadership, for example in fuel-cell technology, may simply be forfeited.

It was clear from our research that, notwithstanding the uncertainties and challenges, business tends to hold the following viewpoints:

1. Market-based approaches for reducing greenhouse gas emissions and developing and applying new products and services remain the most promising instruments for achieving cost-effective change (as opposed to direct regulation or taxation).

2. There is no technological or public policy "silver bullet." A combination of new energy sources and incisive public policy interventions will be required to fuel the future in a sustainable manner. Meanwhile, there is a significant opportunity for companies that develop capabilities in energy efficiency, new products and services, and emissions trading to gain competitive advantage.

3. There will be a very large demand for energy and energy services in the developing world, especially India and China. Business, in partnership with governments and others, has an opportunity to steer both the developed and the developing world in a more sustainable direction under the right economic, social, and political conditions.

A final point that emerged in our research relates to the topics of leadership and corporate culture or values. When we asked our interviewees how they were achieving breakthroughs in energy efficiency or the introduction of new technologies, the most frequently cited factor was leadership at every level of the organization: champions at the top, in middle management, and on the shop floor. Interviewees also emphasized the importance of a culture of commitment to a "triple bottom line" vision, manifested as planning for the longer term, responding to external drivers rather than waiting for regulations, and engagement with stakeholders. Direct success factors for energy efficiency and innovation were certainly important: for example, reducing costs, overcoming technological and institutional barriers, and actively replacing inefficient capital stock. But we were left with a clear picture of the deeper reasons for companies achieving significant progress both today and tomorrow, and those reasons involve the desire on the part of business organizations to lead, and to do so within a framework of sustainability values.

And so the final challenge is a social one. How can society — as represented by individuals, governments, and civil-society organizations around the world — help create the conditions in which more businesses lead, and are rewarded for leading in sustainable energy solutions? Surely, there is a special opportunity here for Canadian social institutions and Canadian business to act as role models in creating effective partnerships to deliver international leadership in sustainably fueling the future. Based on our research, we believe that the potential for such leadership exists.

A HISTORY OF ENERGY (CONTINUED)

1939 The Austrian physicist Lise Meitner, along with her nephew Otto Frisch, proves that splitting the nucleus of the uranium atom is possible and coins the term "fission" in the English magazine *Nature*. When Frisch is later awarded the Nobel Prize in chemistry for his role in this discovery, Meitner will be overlooked.

1941 The world's first large-scale wind turbine is erected on Grandpa's Knob in Vermont. The 1.25-megawatt machine will sell power to the grid until 1945. The United States government secretly establishes the Manhattan Project. Overseen by J. Robert Oppenheimer, its objective is to build an atomic weapon before the Germans do.

1942 An atomic pile — the world's first nuclear reactor — is demonstrated by the Italian-born physicist Enrico Fermi on a volleyball field beneath Chicago Stadium. He is later to win a Nobel Prize for his work on nuclear reactions.

1943–45 Canada's first offshore oil well is drilled by Mobil from an artificial platform off the coast of Prince Edward Island. No commercial oil and gas are found, despite a drilling depth of 4,500 metres.

1947 Imperial Oil strikes oil at Leduc, Alberta. This marks the commencement of a veritable oil boom in Canada.

1948 The first plutonium production reactor in the Soviet Union comes on-line.

1950s Oil replaces coal as the world's main fuel.

1952 U.S. President Dwight Eisenhower breaks ground for the world's first commercial nuclear power plant, at Shippingport, Pennsylvania.

1954 Bell Lab scientists Calvin Fuller and Daryl Chapin develop photovoltaic (PV) cells made of phosphorous and boron and capable of a 6-percent energy-conversion efficiency when used in direct sunlight. This level of efficiency opens up the possibility of solar cells running everyday electrical appliances. The first real application of PV cells will debut four years later when they power the radio transmitter on the American satellite *Vanguard 1*.

▼ CONTINUED ON PAGE 306

11.

LIFE AFTER OIL: A NIGERIAN EXPERIMENT

KEN WIWA

EDITORS' INTRODUCTION

On November 10, 1995, the military rulers of Nigeria ordered the execution of the writer and activist Ken Saro-Wiwa. He was taken to a prison in the city of Port Harcourt, where, along with eight other activists, he was hanged. Four hours later they were buried together in an unmarked grave.

Saro-Wiwa's unjust and shameful execution occurred as a result of his protests against the activities of multinational oil companies — particularly Royal Dutch/Shell — which had extracted billions of dollars' worth of oil from the Ogoniland, a once-fertile corner of Nigeria. The Ogonis had seen little of the economic benefits, while the pollution from the oil fields were linked to agriculture failure and social decay. The true cost of oil, for the people of Ogoniland, was far too high a price to pay. At his defence, Saro-Wiwa made the following statement:

> We all stand before history. I am a man of peace, of ideas. Appalled by the denigrating poverty of my people who live on a richly endowed land, distressed by their political marginalization and economic strangulation, angered by the devastation of their land, their ultimate heritage, anxious to preserve their right to life and to a decent living, and determined to usher to this country as a whole a fair and just democratic system which protects everyone and every ethnic group and

gives us all a valid claim to human civilization, I have devoted my intellectual and material resources, my very life, to a cause in which I have total belief and from which I cannot be blackmailed or intimidated . . .

Ken Saro-Wiwa's son Ken Wiwa spent most of his life running from his father's legacy, not wanting to fight his father's battles. But as his father's situation became dire, Wiwa took the lead in the movement to try to get him released from prison. Since his father's death Wiwa and his family have moved to Canada, where most people know him as the man who writes a column for the *Globe and Mail* called "Without Borders." What they don't know is that every three months Wiwa goes back to Nigeria and Ogoniland to continue working on his father's dream of building a better society.

More than most people, Wiwa has been directly affected by the global quest for energy, and we wanted to know whether the situation in Ogoniland has actually improved since the late 1990s. After all, this should be a test case of the kind of social and technical ingenuity we've heard about so much in this book. Have the oil companies actually improved their business practices? Is it possible to supply enough ingenuity to a place like Ogoniland in order to balance both the business interests in extracting oil and the civic interests in building a healthy society?

Wiwa relates his answer to those questions and how it led him to start an ingenuity project of his own, one that might allow Nigeria to escape the trap of over-reliance on oil.

IF NIGERIA DID NOT EXIST you could not invent it. Imagine a country that is the sixth-largest oil producer but has to import gasoline and is one of the poorest in the world. A country that provides constant electricity to most of West Africa but suffers from frequent power outages. A country whose scientists, writers, and engineers have won the highest accolades around the world but is a byword for corruption and systematic inefficiency. Nigeria is a country that defies logic; it is a country where, as my father once lamented, "The only wrongdoers are those who do no wrong."

How can a country with such abundance in human and natural resources have so squandered its assets?

That is the question that 120 million increasingly disillusioned Nigerians have been asking themselves for the past forty-three years, since the country won independence from Britain in 1960. It is the question that animated my father's life and contributed to his murder in 1995. It is a question that inspires me to try to find a way out of the cycle of corruption and poverty that threatens to annihilate future generations of Nigerians.

I am an Ogoni. That means I am one of an estimated 500,000 people, a minority ethnic group in Nigeria's mosaic of 400 or so nations. We live on 404 square miles of an oil-rich and fertile plateau of the Niger River Delta in southern Nigeria.

Oil was discovered in Nigeria in 1956 and in Ogoni two years later. Since then an estimated 900 million barrels of oil have been pumped out of my community. But instead of benefiting from this resource, the presence of oil companies has proved to be a curse for the people and a blight on the land. A region that ought to be as rich as a small gulf state remains underdeveloped. Pipe-borne water is virtually non-existent, the electricity supply is barely functional, schools and health services are underfunded and scarce. The rudimentary provision of social services is a poor return in a community that has provided an estimated US$30 billion of oil to Nigeria's treasury.[1]

And to compound my people's misery, the activities of an unregulated, largely unaccountable oil industry has strained the fragile ecosystem of the community. For over forty years of prospecting the oil industry has profited very well from Ogoni but paid scant attention to the impact of their operations in and on my community.

Up until the end of his life my father campaigned vigorously to bring our suffering to national and international attention, pointing out that "thirty-five years of reckless oil exploration by multinational oil companies has left the Ogoni environment completely devastated." When he spoke of Ogoni, he was also speaking of Nigeria. In his mind Ogoni was Nigeria writ large, a story of squandered riches and wasted potential.

In 1990 he formed the Movement for the Survival of the Ogoni People (MOSOP). In his vision of MOSOP Ken Saro-Wiwa saw a non-violent grassroots organization that would not only sensitize our people to what was happening to our land, people, and culture but would also mobilize the community to stand up for its rights. So successful was his mobilization of the community that Shell pulled out of Ogoni and has not been able to return since. The military government of the time responded by harassing community leaders and eventually arresting and trying my father on trumped-up murder charges. Despite global protests and diplomatic intervention, he was hanged on November 10, 1995.

The military dictatorship wanted to teach the Ogoni and other oil communities in the Niger Delta a lesson. My father was innocent and his only crime was in speaking out against the petro-politics that continue to undermine the nation: oil accounts for 90 percent of Nigeria's export revenues, and the billions of dollars that accrue from oil are what attract corrupt soldiers and politicians to power in Nigeria.

The Nigerian political economy is dysfunctional as a consequence of the peculiar land-use and mineral-rights legislation of the country, which vests all land and the rents and royalties within that land in the hands of the central government. With a centralized economic and political system, the struggle for power and patronage is the substance of politics in Nigeria. And in that

AS WELL-PLACED INDIVIDUALS CONTINUE TO ACCUMULATE UNACCOUNTABLE WEALTH, THE COUNTRY CONTINUES TO SLIDE INTO SOCIAL CHAOS, ITS HEALTH, EDUCATIONAL, AND ECOLOGICAL ASSETS WASTED AND UNDERCAPITALIZED.

struggle minority ethnic groups like the Ogoni are disenfranchised, as are the increasingly impoverished masses who have no voice or representation in the fraudulent politricks of Nigeria's venal and often unaccountable elites. Protest is met with indifference or repression while billions of dollars are spirited away by a handful of power brokers in Nigeria's tortuous experiment with democracy. As much as 100,000 barrels of oil a day[2] is smuggled out of the country, and it is said that the amount of money held by Nigerian individuals in foreign bank accounts is equal to

Nigeria's external debts of US$30 billion. As well-placed individuals continue to accumulate unaccountable wealth, the country continues to slide into social chaos, its health, educational, and ecological assets wasted and undercapitalized.

I am writing this in my father's study in the family home in the oil city of Port Harcourt. It was in here that my father formulated many of the ideas that animated our community to rise up and protest against the economic crimes being committed against us. He was often in here, late at night, writing in longhand under the bright light of a neon strip light. I, a generation later, sit here by candlelight, typing this on my laptop.

As new technology opens up new horizons for humankind, we in Nigeria are slipping back into the Dark Ages. Nigeria is like a crazy quilt of the old and the new; a baroque mix of the latest consumer goods littered around a landscape of poverty. While the elites accumulate unimaginable wealth, feeding off the country's assets, millions of Nigerians have to survive on the scraps.

For some it is a hopeless task, and many dream of escaping to North America and Europe. Some have resigned themselves to a life of prostitution and crime. As corruption continues to devour health and education budgets, a generation of Nigerians have no option but to find ingenious ways to make a living. Scam letters, faxes, and e-mails have become one of Nigeria's biggest and most infamous exports. The culture of preying on the unsuspecting and greedy is a direct consequence of the authorities' failure to build an economy that can provide sustainable employment for its citizens; Nigeria's oil wealth, the US$15 billion it accrues each year from the commodity, insulates and blinds them to the problems in the country. Nigeria has earned some US$300 billion from oil in the last thirty years but little of it has been invested in infrastructure and capital projects. As my father used to say, Nigeria's greatest crime is the failure to think.

By any account Nigeria should be bankrupt, but it is the

ingenuity, the perseverance, the sheer bloody mindedness of its disenfranchised masses that keeps the country solvent. When all the politicians have looted the treasury, all that remains is for most of Nigeria's 120 million people to eke out a living from the scraps that trickle down from the commanding heights of the economy.

For many years I was happy and lucky enough to have escaped the deprivations of living in Nigeria. Although I was born and grew up here, I was mostly educated abroad. I was content to live a life of relative comfort until my father's political conscience sucked me into the vortex of Nigeria.

After a long process of self-interrogation I arrived at a conscious decision to find my own answers to the questions that my father lived and died for. I began to make trips back to Nigeria, at first to clear up the outstanding details of my father's death, but increasingly to make my own contribution to solving Nigeria's problems.

I was lucky that I was able to come back after the sixteen years of brutal military dictatorships had ended in a transition to

NEW TECHNOLOGY, WITH ALL ITS ATTENDANT PITFALLS AND BENEFITS, IS STARTING TO ALTER THE SOCIAL AND ECONOMIC DYNAMICS OF THIS COUNTRY.

civilian rule in 1999. Although democracy has not brought about many of the expected benefits, it has brought some changes, especially in the telecommunications sector.

In 1999, a country that had been in the nineteenth century in terms of communications began a headlong rush into the twenty-first. Within two years a country that had only 600,000 land phone lines suddenly had 2 million subscribers to its two cellular companies. The Internet arrived in Nigeria and has been embraced by a generation starved of entertainment and desperate for links to the outside world.

New technology, with all its attendant pitfalls and benefits, is starting to alter the social and economic dynamics of this country. Suddenly Nigerians can connect with each other across the nation and around the world, where 10 million live in the diaspora. Of course, it means that we can now send more scam letters and mass e-mails, copy music and distribute it without fear of prosecution, and divert satellite signals to makeshift dishes in isolated rural areas. But it also means that a country that has produced Philip Emeagwalli — the man credited with inventing parallel computing — can harness its intellectual assets in the new economy.

Anyone who knows Nigeria will not be surprised at Nigerians' innate ability and ingenuity with modern technology. As I observed these developments over two to three years of travelling back and forth between Canada and Nigeria, I began, like many entrepreneurs, to think of the many ways Nigeria's ingenuity could be harnessed to find solutions to the country's dysfunctions.

The result is suanu.com, a pilot project to assess the viability of a communications network to serve as a platform and virtual infrastructure for sustainable local enterprises in Ogoni, Nigeria. It is an attempt to find a way to link my community to the outside world, using new technology and a business model that can be profitable yet has a strong social mission to provide services in health, distance learning, and running agricultural co-operatives. The big idea is that businesses can fill the void left by inefficient and corrupt governments.

In the year since I conceived of the project I have encountered the usual obstacles that stand in the way of any initiative in Nigeria, the most obvious being financing. You discover that the technology gap between the North and South is a cost issue — Internet access is roughly ten times as expensive in Africa as in the rest of the world. Moreover, since banks charge as much as 40 percent interest on loans, one invariably has to look abroad to find ways to raise capital. And who wants to put money into Nigeria, whose reputation for corruption and malpractice and daily stories of instability and civil unrest discourage all but the most intrepid investor?

You despair at the prospect of having to invest a third of start-up costs and as much as 20 percent of running costs in fuel and generators to ensure a constant source of electricity. It seems absolutely criminal that so much capital has to be invested in energy for a business that is located eight kilometres from two oil refineries, in a country that produces 2 million barrels of oil a day.

But as my father used to say, every problem contains a solution. And that's why I believe that fuel-cell technology may just be the answer to the problems that bedevil my community and country. The technology offers the possibility not only of a cleaner environment, but of forcing Nigeria and its political elites to develop alternative sources of income and to rely on our intellectual rather than natural assets.

As I write, fuel-cell technology is virtually non-existent in Nigeria, which is still wedded to cheap fuel. But those of us who have fought the oil cartels and mafia that have ruined Nigeria suspect that this country will never reach its full potential as long as the world remains dependent on fossil fuels. As my father understood, to fight the cartel is futile, but the lessons learned in that fight, if learned well and properly applied, will serve disadvantaged groups and communities in Nigeria. In short we must challenge injustice with our intellectual resources.

Clean technologies and the knowledge economy offer the distant possibility of salvation, and I'm increasingly taken by the idea, delighted at the irony of running a new technology business powered by clean energy in a community that has been scarred by a history of fossil fuel extraction.

It would give me immense satisfaction to pull this off, to prove what my father always insisted: that only the acquisition and use of knowledge will ultimately save Ogoni from our predicament in Nigeria. The knowledge economy offers us the prospect of building communities, countries, and a world that is cleaner and less reliant on dirty fuels, a world that forces everyone to manage and invest its intellectual and human capital for sustainable development.

For me suanu.com is the beginning of that dream, though it may be a distant dream, perhaps no more than a stone cast into the pond of Nigeria's problems but you hope the ripple effects will be wide-reaching. The name I have chosen for the project is, like all African names, significant: in my language *suanu* means knowledge.

A HISTORY OF ENERGY (CONTINUED)

1955 The Soviet Union develops its first nuclear power-generating station at Obninsk.

1962 Canada's first nuclear reactor goes on-line at Rolphton, Ontario.

1968 While working as a machinist for an incineration company, Rufus Stokes receives a patent for an air-purification device that will reduce the gas and ash emissions from furnace and power-plant smokestacks. The "clean air machine" as he dubs it will not only reduce health risks to humans but will also improve the appearance of city buildings that are exposed to pollution.

1973 The OPEC oil embargo sets off the first global energy crisis. The price of a barrel of oil increases by 70 percent.

1977 In a bid to move the United States away from dependency on foreign oil, U.S. President Jimmy Carter announces incentives for renewable energy systems and installs solar panels on the White House.

1979 The Three Mile Island reactor has a partial meltdown.

1980 The Florida car company Vanguard markets the first electric vehicle. It can travel up to 64 kilometres at 80 km/h and costs $3,500. General Motors will be the first major car company to follow suit with its Impact electric vehicle in 1991. Despite excitement around electric vehicles, the limited distance between powering will limit their commercial success.

1984 The Jeep Cherokee, the first mass-production SUV, debuts. Solar power in the United States drops to US$10 per watt, on its way to US$5 per watt by 1992.

1986 The Chernobyl nuclear power station explodes, killing thirty-one people instantly and releasing a radioactive cloud.

▼ CONTINUED ON PAGE 322

12.

IDEAS FOR THE HOME FRONT

AVI FRIEDMAN

EDITORS' INTRODUCTION

No one knew what to make of the house. It was a svelte and clean-looking thing, as if an old Victorian rowhouse had been tossed in a dryer and shrunk. Built in 1990 right in the centre of the main campus of McGill University, the house was meant to make a statement of some kind, but we couldn't figure out what that statement was. There were whispers about it, of course. It was an affordable house, someone said. Even arts students might one day be able to buy one. That seemed implausible. There was no such thing as an affordable, roomy house. Even arts students knew that.

Avi Friedman and Witold Rybczynski proved us all wrong. The two McGill architecture professors were the designers of the structure they called the Grow Home. It was an exercise in ingenuity, their way of solving some of the most intractable urban problems, including the increasing cost of housing and suburban sprawl. Is it possible to build a small, efficient, affordable house and still make it livable? Indeed it is.

Only about five metres (fifteen feet) wide, the Grow Home offered a mere 93 square metres (1,000 square feet) of living space. But when you wandered through it, the house felt airy, with open ceilings and simple techniques for taking advantage of space. The design used very simple building techniques to achieve a price per square foot of less than $40. That meant it was possible to build a house where four people could live comfortably for a total cost of $40,000.

Their idea immediately garnered widespread interest, and today Friedman conservatively estimates that over 10,000 Grow Homes have been built around the world. Many of those have been in Montreal, but places as far away as China and Mexico have looked seriously at the concept of this ingenious house. More than 400 newspapers and magazines across the globe have featured the Grow Home, and it has earned Friedman numerous awards, including the World Habitat Award of the United Nations, which he accepted at a ceremony in China on World Habitat Day (October 4) in 1999.

Friedman's work on the Grow Home has led him to think seriously about innovations in energy efficiency, which is why after seeing his work on campus thirteen years ago, we believed it was important to hear from him again. In this chapter, Professor Friedman brings his ingenuity and practical solutions to our homes, looking at simple ways that homeowners can reduce their electricity or home heating and cooling bills, without sacrificing comfort and without spending a small fortune.

With the debate over the Kyoto Protocol receiving considerable media attention over the last year, we've heard a lot of discussion about its impact on the business sector, the energy sector, and various governments. But what will this mean to the average consumer? How will you be affected, and how might you reconfigure your behaviour to reduce your personal "energy footprint"? The fortunate truth is that, for most of us, there are a number of places in our lives where we can make a difference to the way we consume energy. And foremost of those is the type of house we live in. You could say that energy efficiency — like many other things — begins at home.

Domestic energy consumption currently accounts for 12 percent of Canada's and most other developed nations' total energy use.[1] As a result, there may be no single bigger way to reduce carbon dioxide (CO_2) emissions than to reconsider how we use energy in our houses — and how we might use it more efficiently. And with the increasing costs of home energy and heating, there's a real incentive — and a potential financial payback — for those homeowners who consider ways to use energy more efficiently. In this chapter, I'll look at several economical ways that the average homeowner could reduce their home's energy output and help lessen both home heating bills and greenhouse gas emissions.

TAPPING THE SUN

One of the best energy-saving measures that you may wish to consider is achieved by using solar energy. Not many people associate northern regions with sunny days. Most tend to picture the wintertime sun rising and shining in warmer places like California or Florida. Surprisingly, despite the shortness of our winter days, many of them are sunny. Yet we fail to take full advantage of the cheapest, cleanest energy source of all — the sun.

Solar power design for buildings can be divided into two categories: active and passive. Active solar design refers to strategies that incorporate mechanical systems in a building. These could be photovoltaic cells, which — much like those on our calculators — generate energy when exposed to the sun. Solar water heaters are also included in this category, although they typically use solar thermal energy, using the sun's rays to generate heat energy rather than electricity. Active solar design has very real potential for increasing energy efficiency, and is being integrated into buildings across North America on an ever-greater scale. And with "net metering" — the right to charge back to the energy utility any excess energy a house might generate at off-peak times — already available in many locations across North America, active solar will be of increasing interest to builders and homeowners. However, active solar is discussed in more detail earlier in this book, and has a capital cost that puts it out of reach of some consumers. Therefore, I'll focus here on the potential of passive solar design.

Passive solar design strategies regard a building's location, its configuration, and its method of construction as key factors contributing to energy gain.[2] Passive solar is easier to implement than active solar if a common sense approach to the location and design of the house is followed.

At the beginning of a good passive solar design, you should ensure maximum building exposure to the sun. Most streets, and consequently most houses, are oriented either north-south or east-west. The ideal orientation for a house will be north-south, with the widest façade facing south. Location on an east-west axis can also work, as long as the southern exposure is not shaded by adjacent buildings. Once the location of the building has been decided and the design has begun, more openings should be arranged on the southern exposure and fewer of them on the northern exposure. Gain will depend primarily on letting the sun shine through the windows and minimizing opportunities for its energy to escape through the colder north-side openings.[3] Therefore, an important consideration will be to select appropriate windows. As I'll discuss later, selecting well-built, energy-efficient windows may prove to be a good long-term investment in general, but even more so when passive solar design is applied.

You should also consider the effect of the summer sun. The welcome winter sun may cause overheating in summertime. The inclusion of appropriate devices, like awnings or shutters designed to follow the higher summer sun angle, will keep a room from becoming a sauna. Once sunlight has penetrated a room, its energy needs to be kept there. This can be accomplished by building a "thermal mass" — a wall located in the path of the sunlight where it may absorb the solar energy. Made of heat-absorbent materials like masonry, this wall will release heat after the sun sets.

Another essential consideration involves the placement of rooms within the house. Common sense has it that rooms you use mainly in the daytime, like the kitchen, living room, and family room, should be placed where the sun shines, to maximize the advantage of light and warmth during the day, when these rooms are being used. Rooms you use at night and purely service rooms, like bedrooms and the laundry room, can be placed against the northern wall, since their peak hours of use are after sunset. Another design factor concerns the conceptual approach to the use of the interior. Creating an open plan that permits heat to

TAPPING THE SUN AND DESIGNING FOR PASSIVE SOLAR GAIN WON'T ELIMINATE OUR RELIANCE ON OTHER SOURCES OF ENERGY, BUT THEY WILL REDUCE IT.

travel easily from room to room and from floor to floor will be a preferred approach. Since heat rises, having open stairs and avoiding skylights from which heat might escape will benefit the home's energy management.

Tapping the sun and designing for passive solar gain won't eliminate our reliance on other sources of energy, but they will reduce it. What we stand to gain is an endless source of clean energy and comfortable homes. In the past half-century, we've relied on a limited number of energy sources, with oil, coal, hydro, and nuclear among the main ones. As the world and individual homeowners attempt to contribute their share to lowering greenhouse gas emissions, we must explore other alternatives, some of which are clean and free. One such alternative is geothermal energy.

GEOTHERMAL ENERGY

Geothermal energy technologies use heat from the Earth for applications that range from powering a heat pump for a single home to large-scale electrical power production. In most locations, the upper three metres of the Earth's surface maintains a constant temperature of between 10°C and 16°C. Geothermal heat pumps can tap into this reserve to heat or cool homes.[4]

The system consists of a heat pump, an air-delivery system (the ductwork), and a heat exchanger. A network of pipes is buried in the shallow ground near the building. Alternatively, where space is limited — as is often the case in a city, for example — the pipe may be buried vertically in the ground; closed loops of pipes are inserted into a borehole with a diameter of 10 centimetres and to a depth of 25 to 90 metres. During the winter the heat pump removes heat from the heat exchanger and pumps it into the indoor air-delivery system. During the summer the process is

AND WHEREAS THE COST OF CONVENTIONAL HEATING OPTIONS — ESPECIALLY NATURAL GAS — IS CONTINUALLY ON THE RISE, THE GEOTHERMAL CUSTOMER IS PROTECTED FROM PRICE INCREASES BECAUSE GEOTHERMAL ENERGY IS PROVIDED FOR FREE BY THE EARTH AND THEREFORE ITS OVERALL COST STAYS CONSTANT.

reversed, and the heat pump moves heat from the indoor air into the heat exchanger. The heat removed from the indoor air during the summer can also be used to provide a free source of hot water.

A slightly different system exists for homes located next to a lake or pond. In this case the coiled pipes are installed at the bottom of the water source and the heat is transferred to or from the water. This system makes use of the fact that heat transfer is much more efficient through water than through soil. Another variation of the geothermal energy source is the vertical system, which is useful when the land area is limited.

In the past decade geothermal energy has become popular with homeowners. As compliance with the Kyoto Protocol looms and countries soul-search, geothermal sources offer both reductions in CO_2 emissions and cost savings. Homeowners and architects who introduced these options to their clients suggest that the costs compare favourably with those of other alternatives. In fact, geothermal is the hands-down winner in terms of operating costs. For a 230-square-metre (2,500-square-foot) house in Toronto, Ontario, annual heating costs for natural gas are around $2,000; for oil they're slightly higher, around $2,300; and for propane they're around $3,400. But with geothermal, those operating costs are in the range of a mere $700 a year. The catch is the capital costs required for a geothermal system, which are around $10,000 to $15,000 for the average home.[5] However, when you add this all up, the payback time is a very reasonable eight years, far superior to other green-energy options like active solar. And whereas the cost of conventional heating options — especially natural gas — is continually on the rise, the geothermal customer is protected from price increases because geothermal energy is provided for free by the Earth and therefore its overall cost stays constant. Moreover, once the initial investment is paid, the homeowner gets an almost free supply of energy, with no more maintenance or servicing than might be required by a conventional furnace and air conditioner.

Geothermal energy sources are now also being used to generate electricity. Since the first generator was introduced in Larderello, Italy, in 1904, their use has grown worldwide. The Geothermal Education Office in the United States suggests that 2,700 megawatts of electricity are produced there from geothermal sources, sparing the country from importing some sixty million barrels of oil each year.

OTHER RENEWABLE ENERGY POSSIBILITIES

Some other alternative energy sources are also being explored. Their machinery cannot be buried in the backyard like it can for geothermal sources, but they hold significant potential to lower both CO_2 emissions and heating bills. Landfill bio-gas is one example. It is generated from household wastes and is produced when these products are fermented in the absence of oxygen. It can provide a cheap source of energy primarily in rural areas.[6]

Another example is tidal energy, which can be generated in coastal locations from the twice-daily ebb and flow of the tides. As the tide rises, water is allowed to flow through gates in a dam and fill the basin behind. At high tide the gates are closed, and as the tide falls, the water in the basin is retained behind the dam. Once a sufficient head of water is built up, the water behind the dam is released and drives a generator to produce electricity.[7]

DESIGN PRINCIPLES FOR ENERGY EFFICIENCY

In addition to these renewable-energy possibilities, there are a number of fairly simple design features that can be integrated into many homes to make them more energy efficient. In the past, the recommended strategy for energy efficiency was to construct super-insulated homes. Their elevated costs explain why they were not popular with the public. But today I think there's another, more viable approach. I'd therefore like to propose a "common sense" approach to better manage energy consumption for the home that takes advantage of relatively simple and cost-effective solutions many homeowners could employ.

If you wish to minimize heat losses, you can use the highest level of energy-efficient construction materials, components, and techniques. The problem is, these options are available to only a limited segment of the population due to their high construction costs. So in most typical homes, the best approach — the approach that is most affordable for the average homeowner — focuses on optimization rather than maximization. The savings come from the use of simple, common sense solutions rather than from sophisticated construction materials and methods. Through efficient planning principles and appropriate construction assemblies, you can reduce heat losses by half without compromising your living comfort.

The easiest way to reduce your energy cost — and save expenses in addition while simultaneously reducing your energy footprint — is to look at reducing the size of your home (Figure 1). While many people are loath to do this, in fact a comfortable living accommodation can be provided within a relatively small floor area of 100–200 square metres (1,080–2,160 square feet). This can provide you with energy savings of as much as 12 percent compared to an average starter home of 120 square metres (1,300 square feet).[8] And given the shrinking size of the average North American household, most people can achieve this reduction without compromising their living comfort. As a result, you can eliminate wasted floor area while reducing the cost of heating unused space.

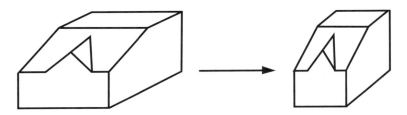

Figure 1: Size reduction

Another effective way of reducing your heating costs at the design stage is by simplifying the building floor plan (Figure 2). A rectangular plan, for example, has about 7 percent less perimeter than an L-shaped plan of the same area, 17 percent less than a U-shaped plan, and 24 percent less than an H-shaped plan. By reducing your building's perimeter, you'll gain in energy savings both directly and indirectly, as a result of reducing the exterior wall area, the infiltration at the joint between the foundation wall and the exterior wall, the number or size of windows, and the number of corners.

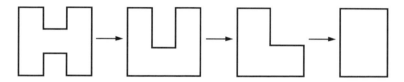

Figure 2: Plan simplification (not to scale)

Stacking floors on two levels may provide you with additional energy savings (Figure 3). Although the exterior wall area is increased when floor space is divided on two levels, the potential additional heat losses are usually avoided through a reduction in roof and basement areas from which most of the heat tends to escape. Infiltration losses at the joint between the foundation and the exterior wall are also lowered since the overall building perimeter is reduced, and the basement requires fewer windows from which heat can escape.

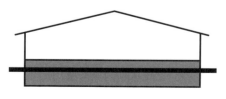

One Floor (bungalow)

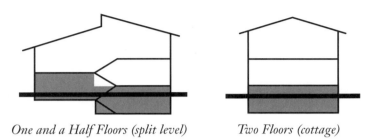

One and a Half Floors (split level)   *Two Floors (cottage)*

Figure 3: Floor stacking

The greatest advantage when your home is built taller and the floors are stacked, however, is that adjacent houses can be joined as semi-detached units or townhouses (Figure 4). This provides significant energy savings through a major reduction in exposed wall area, particularly when the front of the house is narrow.[9]

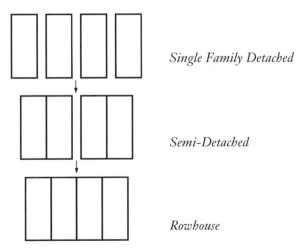

Single Family Detached

Semi-Detached

Rowhouse

Figure 4: Joining of units

WINDOWS

One of the most effective ways you can control heat losses is through the careful selection of windows. There are three aspects of windows to consider: the type of window, the frame material, and the type of glazing.

Whether you're considering a window for a new home or replacing existing ones, be aware that windows that require sliding assemblies are generally less efficient than those that require a pivotal mechanism. Fewer operable parts lead to greater efficiency. The type and material of the window frame also has an effect on the window's overall thermal performance. Wood frames are usually more effective insulators, followed by vinyl, thermally broken metal (those with rubber gaskets), and metal frames. In addition, you may want to make sure that all the units have weather-strips around the moving parts.

The other important consideration in the selection of windows is the type of glazing. The upgrade from single-glazed units to double glazing promises the highest savings. The addition of low-emissivity coatings that are placed on the glass can also increase the thermal performance of the window.

BASEMENT

If you're considering whether or not to include a basement in your newly designed house, consider the climate in which it's being built. In cold climates, building codes may require that the foundation walls be constructed to a depth of 125 cm to 140 cm below grade to ensure that the footings rest below the frost line. If this is the case, a basement could be provided at a small marginal cost, increasing the floor space by one-third. For row houses, heat losses would be increased by around 11 percent. Because the heating costs for this type of housing are relatively low, this increase would normally represent a small annual fee. Considering that a basement would increase the total living area by one-third, the added expense might prove worthwhile.[10]

CONCLUSION

The good news is that smart homeowners can do their part to become more energy efficient, without breaking the bank. In fact, done wisely, increasing your home's energy efficiency can lead to real savings over time, especially with natural gas and other heating prices currently on the rise. A series of adaptations or optimizations is all that's required to improve the energy efficiency of your house. And if we turned to these techniques on a larger scale, the potential impact and energy savings would be dramatic.

A HISTORY OF ENERGY (CONTINUED)

1993 | Ballard Power Systems demonstrates three fuel-cell buses in Chicago. These are the first vehicles to use a modern fuel cell.

1999 | DaimlerChrysler unveils its NECAR 4 (short for New Electric CAR). It is the first commercial fuel-cell-powered car. Ballard Power Systems introduces six hydrogen-powered buses to Chicago and Vancouver. Wind power is the fastest-growing energy source in the world, with average annual growth of 25 percent and a 90-percent drop in price since 1990.

2000 | The Ontario Medical Association releases a study estimating that air pollution kills 1,900 people in Ontario each year.

2001 | Iceland unveils plans for a fleet of hydrogen buses, a first step towards developing the first hydrogen economy. The U.S. Environmental Protection Agency reports that the average fuel economy of all car and truck models sold in the United States has fallen to 20.4 miles per gallon (11.5 L/100 km), the lowest level in two decades, largely due to the popularity of SUVs, light trucks, and minivans.

2002 | An American senate amendment requiring cars to become 50 percent more efficient over the next thirteen years is rejected.

2003 | Two solar water heaters are installed on the White House. China's massive Three Gorges dam is completed. The world's first commercial hydrogen filling station opens in Iceland. There is a projected record in drilling activity for natural gas in Western Canada, spurred by continent-wide high prices for the fuel. Plans are announced for the 1,400-kilometre Mackenzie Valley natural gas pipeline, North America's first twenty-first-century megaproject.

13.

GENERATION 2023: TALKING WITH THE FUTURE

MICHAEL BROWN

EDITORS' INTRODUCTION

If you attended the 2003 Hydrogen Fuel Cells Conference in Vancouver, you'd have been forgiven for thinking that the battle over energy would have a happy ending. Standing in the foyer at the Westin Bayshore Resort & Marina — shoulder to shoulder with the brightest minds of the hydrogen economy — you could see some of nature's greatest sights all around you. In one direction you could see the mountains in the distance and in the other you could see the ocean undulating just past the tranquil waters of Coal Harbour. And a short walk from the conference site would take you to the towering redwoods of Stanley Park. If ever there was a picture of the future we all want, you could see it out those windows.

Vancouver has become the unofficial hydrogen capital of the world. This is partly because some of the largest and best-known hydrogen companies — including Ballard Power and General Hydrogen — have chosen the Vancouver area as their base of operations. But it's also partly because the city seems to embody the hopeful optimism surrounding the concept of a hydrogen economy. Vancouver is a city perpetually on the edge of becoming the next big thing, a careful balance of human, commercial, and environmental interests. But as with the hydrogen economy, the ideal vision of Vancouver is still shimmering somewhere in the future, a dream that has yet to reach its full potential.

One of the people you might have met at the conference was Mike Brown, an elder statesman of the Vancouver fuel-cell industry. As president of Ventures West, B.C.'s leading venture capital firm, he spearheaded an investment into an unknown company by the name of Ballard Power. Even though Brown's firm sold his interest in Ballard Power in the 1990s, he retained his interest in power technology, and in fuel cells in particular. Now he is the chairman of Chrysalix, perhaps the only venture capital firm in the world to be exclusively focused on developing hydrogen-based technologies. One of Chrysalix's major efforts is to project into the future, to try to understand the forces and factors that may affect our use of energy in the years to come. Some of this important work informs the following very original look back from our future. Brown's vision of what lies ahead plays an important role in what decisions we're making now and how they're going to affect the generations to come.

NICHOLAS

In May of this year, after I had turned 24 and graduated from UBC with my math degree, my Dad took me to the car store on his sixty-fifth birthday. We test drove a couple of cars with ordinary engines and two fuel-cell cars. I was surprised at how quiet and responsive the fuel-cell cars had become, and I liked all their electric equipment, including built-in Wi-Fi with GlobalNet and remote self-repair. We took the fuel-cell cars up the Cypress Bowl hill, which is just as steep now as I'm sure it used to be, and we were both impressed that we could go 150 kilometres and still have something left. Dad wasn't too sure he wanted to buy a fuel-cell car, but he decided that the risk had now largely disappeared, so we bought a four-door 65-kilowatt Dodge Independent, paying just over $49,000. That was a little bit more than a standard car, but I persuaded Dad that the cheaper maintenance and fuel would more than make up the difference in only a couple of years. And he really liked the smooth performance and the classy shape.

Next day, I proudly took my Grandpa Mike for a ride. I knew he'd once worked in the fuel-cell industry, but I wasn't quite sure what he had done there. He decided that he would drive, even though he was in his mid-eighties and didn't see very well any more. He wanted to drive to Squamish. "Great torque," he said as we blew up the hill on Taylor Way. "Quiet, too. I always knew these things would make you sit up and pay attention. Our big challenge was to get them to work for more than a few months at a time. It took longer than I thought it would, but now consumers finally have a real choice."

"Tell me more."

"Okay, Nicholas, it's quite a story. Expectations were way too high at first. When the century began and you were just learning to walk, people would say we'd have these cars by, oh, 2010. Before then, I had been even more optimistic! I can recall sitting in a dentist's waiting room back in 1986 reading in *Popular Science* magazine that by 1995 fuel cells would be generating electricity all across the United States. In fact, that idea was what first got me interested in the technology.

"Well, 1995 came and went. In 1998 an article in the *Economist* magazine got me quite excited by predicting that by 2016, 50 percent of the cars manufactured would run on fuel cells. That would be 20 million cars, and at $4,000 per engine this would produce gross annual revenues for the industry of $80 *billion*, and by 2020, this growth, together with infrastructure investments, manufacturing plants, and distributed generation systems, would bring in annual revenues of $1 *trillion* or so. I believed every bit of this. I was a fan.

"As it turned out, it was a big deal when in mid-2003 Fuel Cells Canada announced it was getting five Ford Focuses for tests in Vancouver. Honda released five in California, and Toyota also released a few, but theirs had leaks and had to be recalled.

"By 2005, there were finally some fuel-cell products on general release, but they were all in fairly small niche markets. Only a few fuel cells were in operation providing small amounts of electricity, though United Technologies and Fuji and others had been doing beta tests for almost ten years. At the turn of the century, investment analysts had been predicting there would be one million fuel-cell cars on the road by 2010. In fact, nothing really became commercial until around 2012, and even then there were only 50,000 of them around. Only the wealthiest customers or companies could afford them."

"What happened?" I asked.

"No question about it, there was too much hype. But back around the turn of the century there was hype everywhere, especially when the terrorists attacked the United States and everyone got worried about energy security. They loaded this on top of concerns about global warming — you'd have to agree they were right to be worried — and it seemed that fuel cells were the perfect answer because they were so much more efficient, they ran on hydrogen, they produced no carbon dioxide, and so on. In addition, people were getting very upset about the smoggy crud you would see covering Vancouver if you were looking down from Cypress Bowl or that you'd breathe when walking down Broadway. With fuel cells there'd be no nitrogen oxide, Nicholas. No nitrogen oxide means no smog."

"What's nitrogen oxide?" I asked.

"It's a pollutant that comes from gasoline combustion in air. It's a nitrogen compound created by the high burning temperatures, because air is mostly nitrogen. The kind of fuel cell there is in this car operates much cooler, about the boiling point of water, so nitrogen doesn't get involved.

"In 2002 I noticed that analysts were judging the value of Ballard Power based on estimates of cash flow in 2020. That made me stop and think about the forecasts for 2002 that might have been made eighteen years earlier, in 1984. That was before the Internet had reached mainstream consciousness and before global warming was a worry of more than a few oddball scientists. In 1984, Geoffrey Ballard and his colleagues had only started to *think* about fuel cells. So I decided that all those 2020 estimates were just fantasy."

Grandpa turned to me with a sly grin. "But here we are in 2023 and your dad just got his first fuel-cell car. It took a lot of different things coming together to make that possible. And those hybrids sure were tough to beat for a while."

"Yeah," I said. "I remember Mum got a hybrid van in 2005. They were pretty good, weren't they? There's hardly any smog now, and we didn't need fuel-cell cars to do it."

"Well," he said, "a push was started to improve the efficiency of cars — mostly cars, though other energy uses were involved as well. In Europe, they talked about the three-litre car. You didn't hear so much about it in North America. The idea was that if you could get mass transportation based on using only three litres to drive 100 kilometres, or about 90 miles to the gallon, then this might be an answer to the threat of global warming, or at least a temporary answer on the way to the fuel-cell car."

I was a little confused. "But I thought the fuel-cell car was the ultimate answer to coping with global warming."

"Not as long as we got the hydrogen from oil or natural gas. That technology produced carbon dioxide as a residue, but it was a necessary step.

"We used to compare engines using an analysis called 'wells to wheels.' The idea was to measure how much in terms of dollars and greenhouse gases was involved in fueling vehicles, all the way from when the oil or gas came out of the well until the power went through the wheels. We'd break it down into two stages: from the well through the tank up to the point of entry to the engine, and then through the engine to the wheels.

"We did the comparison in two stages because hydrogen is very hard to store, much harder than gasoline or methanol. It's the most dilute substance there is. Think of it this way. A gasoline tank can store several times its own weight in fuel. But the hydrogen in a storage tank, even at the highest compression, weighs only about 6 percent as much as its tank. And customers demand that their cars can travel about 350 kilometres on one tank. Getting enough hydrogen in the space available in a car has always been very hard to do, and it's always been very expensive.

"The first stage, from well to fuel input, costs a lot more with hydrogen than with gasoline. Hydrogen has to be manufactured, and it costs more to store. So hydrogen loses in the first stage. But the use of hydrogen in a fuel cell is so much more efficient than the use of gasoline in an engine that hydrogen always wins in the second stage. Overall, hydrogen came out ahead until gasoline engines reached the efficiency of the three-litre car. Then it was about even. My recollection is that each type got about 40 percent efficiency, which means that about 40 percent of the energy available in the fuel when it came from the well ended up turning the wheels."

I sat up. "So it was the gasoline hybrid that gave us the three-litre car."

"You bet. My old Prius got about 4.7 litres per 100 kilometres, and a later 2004 model got just under four litres. So it was no surprise when Honda announced in 2008 that they had broken the three-litre barrier."

"Well then, why do we have any fuel-cell cars at all, Grandpa? Weren't they expensive?"

"They sure were. There were other problems, too. There weren't any hydrogen filling stations, and fuel-cell reliability just wasn't very good until around 2012. In order to reduce the amount of the expensive platinum catalyst, they had to make the system more sensitive, but when they did that, the membranes would dry out. We also needed breakthroughs in hydrogen storage, and those came slowly.

"But the hybrids actually helped, because if you could hybridize a gas or diesel engine, you should be able to do the same with a fuel cell. So I happen to know, Nicholas, that this car has lithium ion batteries all over the place in it, and they're all charged up either by regeneration as you go down a hill or slow down, or by that small fuel cell. It can run at fairly constant speeds, which helps with longevity and efficiency. This is why you can get such great performance out of such a small fuel cell.

"The other thing that helped was that fuel cells started to be used in other places, like forklifts trucks and standby power. That gave manufacturers some experience." Grandpa paused for a minute, then carried on. "Fuel-cell cars finally came in when global warming became such a big and obvious problem and governments began to push for hydrogen made from renewables. At the same time, about ten years ago, fuel prices began to rise when

WHILE PEOPLE KNEW THAT EVENTUALLY THE FUEL-CELL CAR WOULD SIMPLY BE A BETTER CAR — MORE RELIABLE, FEWER PARTS, MORE CUSTOMER FEATURES, AND SO ON — THE PROBLEM WAS THAT TO GET LOTS SOLD, THE COSTS HAD TO COME DOWN, BUT THE ONLY WAY COSTS COULD COME DOWN WAS IF A LOT OF CARS COULD BE SOLD.

the output of conventional oil and gas reserves declined. That had been predicted as long ago as the 1950s by an oil reservoir engineer from Shell named King Hubbert, but nobody paid much attention then. We've gotten better at making hydrogen now, especially from renewable sources. The government saw that coming and helped with some tax incentives.

"And, of course, there was this difficult problem of how much things cost to make when the volumes are low. While people knew that eventually the fuel-cell car would simply be a better car — more reliable, fewer parts, more customer features, and so on — the problem was that to get lots sold, the costs had to come down, but the only way costs could come down was if a lot of cars could be sold. This was really hard to overcome, but two things happened. The first was that entrepreneurs across North America, Europe, and Asia — especially in Japan — came up with cheaper and better ways to make the systems work, and the second was that there were some early niche markets where the concepts could be tested and the possibility for volume manufacturing scoped out."

As we went up the hill at Furry Creek, we drove by one of the old Urbitaro buses. They had been built in 2010 for the Whistler Winter Olympics, when there was still spring snow in the mountains. I asked Grandpa how those buses got their start.

"Well, Nicholas, that was thanks to the old British Columbia government, which saw that the fuel-cell industry could be a strong part of the provincial economy if things were done right. After all, Ballard Power was started in B.C., which was pure chance because Geoffrey Ballard and his two co-founders decided to live here. Gordon Campbell, the premier of the day, was keen on fuel cells because he understood that a sustainable infrastructure for them would be a permanent job creator. Besides, the feds imposed Kyoto I on us, and this raised awareness of climate change and the need to find alternatives to fossil fuels.

"People in the industry were worried that Ballard might pack its bags and move to Detroit to be closer to DaimlerChrysler and Ford. Campbell called together a committee of people who understood how the fuel-cell industry could come together here — friends of mine like Denis Connor and Ron Britton, and others. They proposed that the government get onside to promote British Columbia as the place for real-life demonstrations of fuel-cell technology. Sort of a 'living lab' of fuel cells. We would do buses and cars, and what we used to call 'distributed generation' before it became the common way to generate electricity. In 2005 these ideas resulted in a deal with the government of China.

"The International Olympic Committee wanted the Olympics to appear environmentally sensitive. Through the local Olympic committee, Premier Campbell got them to agree to buy a fleet of fuel-cell buses. Those became the Urbitaros. I think we made about fifty of them. They cost a fortune, but the tourists loved them and they helped solidify the long-term position of Ballard. At first they had to be supported by another fleet that was never publicized: black vans parked along the highway with mechanics to respond quickly to breakdowns. Just like the Ballard tests in Chicago in the 1990s. By the time the Winter Olympics came along, though, the reliability was as good as the natural gas–fired diesels."

"Why the name 'Urbitaros'?"

"Well, this was a follow-up to Ballard's original bus, which was first demonstrated thirty years ago. I can remember all the controversy we had around the Ballard board room deciding whether to do that. Then, in 2003 and 2004, Daimler put thirty buses powered by Ballard fuel-cell engines in European cities like Madrid and Frankfurt. They called them 'Citaros.' The Urbitaros were a later generation."

"What was the China deal for?"

"Nicholas, it was pretty obvious, back in the late '90s, that what Canadians did about carbon dioxide emissions would be really minor compared to how the Chinese, the Indians, the Indonesians, and the Brazilians handled the problem. Their per-capita use of oil and gas and coal was only about a tenth of ours, and ours was actually pretty stable. You see, Canadians put out only about 2 percent of the global emissions; less now, of course. We once calculated that if those countries I just mentioned got to have just half the per-capita emissions that the Americans did, then annual global carbon dioxide emissions would double, and we'd all be toast. Burnt toast! Developing countries were determined to get to the same standard of living as we had, which would mean the same level of energy consumption.

"When the Americans said they wouldn't participate in Kyoto II either unless the Chinese and other countries started to cut back too, the response was, Hey, this is really a Western plot to prevent us from ever getting to their standard of living. So, just as in Kyoto I, none of them participated.

"Actually, the Chinese and other developing countries were quick to recognize the importance of energy efficiency. It would give a big cost-advantage boost to their manufacturing. Even by 2005, almost one-third of all manufactured goods sold in the United States were made in China. We knew this in the B.C. fuel cell community, and we foresaw a huge export market for our fuel cells in developing countries. But we noticed two obstacles: the Chinese weren't about to buy anything from us unless we proved it worked, and we had to get the price down. Then we had our breakthrough insight: the Chinese weren't about to buy anything from us even if it did work well and the price was right!"

"Now you're talking in riddles. I don't get it."

"The Chinese might license it, but they'd want to make it for themselves and then for the world. Everyone agreed it would be better to put licences in place rather than to have the Chinese simply wait for the original early patents to expire, which would start in 2012.

"Premier Campbell got the feds into the game, and the result was that the Chinese invested about a half a billion dollars in our big demo project. The Urbitaros were a part of it. The feds and the province together put up about the same dollars, so we took a lot of risk. But we got an infrastructure of hydrogen-filling stations while the Chinese got to see what worked and what didn't and got a permanent licence to make fuel cells for Asian markets. We also got a lot of expertise in putting the cars together, in getting the stand-alone fuel cells made and the systems to work, and in diagnosing and servicing them. That expertise now lets us compete internationally with the Chinese and everyone else. Many of our companies have joint manufacturing ventures in China, and some get a permanent royalty stream even though it's not a big percentage of sales.

"The Chinese just dominate the Asian markets, and their products have always worked really well with regular customers, right out of the box. Geoffrey Ballard used to tout the old GM Hy-Wire promo car as being ideal for Asian markets because it would be cheap and people would love to drive it. I don't think anyone much believed him, but Geoffrey was proven right again. In the trial there were more than 500 cars running around Vancouver and Victoria, distributed through an auction run by the British Columbia Automobile Association. I got one of them, and it was great, but not nearly as good as this one you've got."

"It sounds as if demonstration projects have been pretty important."

"They sure are. That's because new technologies face serious chicken-and-egg problems to get accepted. Demonstrations are a way of dealing with these. For instance, we knew it would be really hard to compete with the internal combustion engine for at least three reasons. First, we couldn't bring costs down without volume, but we couldn't get volume without bringing down costs. Second, nobody would build hydrogen-filling stations without customers, but nobody would become a customer without a convenient filling station. Third, to make the vehicles reliable we needed to learn from having them used in brutal day-to-day traffic conditions, but few people would buy vehicles unless they were reliable. Every fuel-cell application has faced exactly these dilemmas, all at the same time. It makes you realize the built-in advantages that protect traditional technologies and discourage new ones.

"This is where the demos came in. We could start working on costs because the government would deliberately pay a lot for the cars to subsidize our learning. The infrastructure got its start because the Urbitaros needed to be fueled. The fuel-cell manufacturers learned huge amounts about reliability. When the trials were first announced, Ballard could get only, oh, maybe 1,500 hours from a stack. The cars needed 5,000 hours. The last Urbitaros manufactured could work reliably for the required 40,000 hours. On top of that, companies like Greenlight were able to install diagnostics. As you know, fuel cells have some features that can sometimes repair themselves on the move, so that through constant contact with GlobalNet, we can be even better assured of their reliability.

"What's this about 5,000 hours? That's only six or seven months. I'd want a car to last longer than that."

"That's driving time. The normal city car operates at about 50 kilometres per hour on average. So in 5,000 hours it travels 250,000 kilometres. Which is pretty good reliability. When fuel-cell manufacturers could count on only one failure in about 5,000 hours, and when they could guess what would fail, they were able

BUT THIS GOT PEOPLE THINKING. THEY SAW THAT THE TOTAL POWER OUTPUT OF THE CARS MANUFACTURED IN JUST ONE YEAR IN THE UNITED STATES IS ABOUT THE SAME AS THE OUTPUT CAPACITY OF ALL THE ELECTRICAL GENERATING STATIONS. MAYBE HOOKUPS TO FUEL-CELL CARS COULD REPLACE THE ELECTRICITY GRID.

to provide a warranty for that period. They would also pay for any extra work needed on other mechanical components if you had to visit the dealer more than once every 50,000 kilometres. Even the best old-style cars can't do that.

"There's something else a fuel-cell car can do that a traditional vehicle can't. Since a fuel cell makes electricity to run motors for the wheels, it's basically a portable electricity generator. Of course, it has to be refueled. But this got people thinking. They saw that the total power output of the cars manufactured in just one year in the United States is about the same as the output capacity of all the electrical generating stations. Maybe hookups to fuel-cell cars could replace the electricity grid. No wonder my old friend Bob Shaw used to call it the 'car appliance.'

"But then economics got in the way. A fuel-cell car engine is expensive, while grid power was available for a few cents a kilowatt-hour. At those prices you couldn't save enough to justify jeopardizing the longevity of your vehicle. Cheaper ways came along to solve this problem. But you can see how fuel-cell technology makes you think outside the box."

We turned off the highway at Squamish and drove down to where the Cheakamus River meets Howe Sound. This is a very windy place — a favourite spot for windsurfers like my cousin Sarah. I was anxious to talk to her about some sustainable development ideas she had, but Grandpa wanted to find Thomas, another cousin of mine, who was an engineer, and who looked after the maintenance on a big windmill situated on the breakwater.

THOMAS AND SARAH

As part of the big B.C. demo project that Grandpa had talked about, Western Wind Farms had imported a huge wind turbine from Europe. Grandpa had some money invested in the project. We walked down the dike, found my cousins Thomas and Sarah, and we all got together at the pub on the main street of Squamish, down by the waterfront.

Grandpa started off the conversation. "Thomas, tell me about that monster out there on the dike. I've heard it's thirty stories high."

"You're right," said Thomas. "The top of the tower is over 120 metres, and each of the blades is 50 metres long. It's a German machine, based on a prototype made about twenty years ago. The Europeans were way ahead of us on this stuff. The Danes were making 15 percent of their electricity from wind even by 2003. They did it to reduce their dependence on imported oil and gas. The national energy company guaranteed Danish farmers that they'd buy all their electrical production at prices that gave the farmers enough prospects to borrow the capital costs for the turbines from their local banks. Very clever plan. The Brits announced in their 2003 Energy White Paper that they'd have to get more power from wind and other renewables because of the natural decline in North Sea oil and gas production. So they went ahead, and we were left behind.

"Anyway, this machine was actually finished in 2009, just in time for the Winter Olympics. It might not look so big from down here, but that nacelle holds a turbine that can produce 4.5 megawatts. In its time it was the highest-output gearless wind turbine. If you think that a normal house would use maybe 3 kilowatts, this would produce enough power for 1,500 homes."

"But it never really was used for homes, was it?" asked Grandpa.

"Well, sometimes. It's a demonstration of how hydrogen can be produced from renewable energy. The electricity output of the turbine was connected to some Stuart electrolyzers in the bus-refueling station over on Mamquam Road. A big problem with wind power is that winds are variable, but here the pressurized hydrogen storage tanks over at the filling station act as a buffer. Once the tanks are filled up to their rated 15,000 pounds pressure, BC Hydro buys the extra power if the wind keeps blowing, so it does end up in homes after all."

Sarah got into the conversation. She's embarked on a career in environmental science, hoping to get some kind of job where she can both make money and put her personal philosophy to practice. She's sort of a 2020s hippy. I admire her, because she absolutely believes there has to be a sea change in how we live. "Footprints on the Earth" is Sarah's favourite saying. Sarah believes this philosophy of sustainability is necessary if we are to have products with few footprints, so she conducts her whole life accordingly.

She asked, "Thomas, what's an electro-whatever you just said"?

Thomas laughed at his cousin's lack of engineering vocabulary. He liked teasing her a bit. "Well, Sarah, do you remember back in grade eleven the science teacher took a beaker of water, inserted a couple of metal rods, which he called electrodes, and ran some direct current to one electrode?"

Sarah looked a bit puzzled. Thomas carried on. "You might remember the end of the experiment, because the current travelled through the water and broke it into its parts: oxygen and hydrogen. Then he lit a match and the whole thing went 'pop!' That 'pop' was the oxygen and the hydrogen getting together again, but the process of separating them this way is electrolysis. While there have been some really interesting breakthroughs in using solar power directly to do the same job, the most basic of the renewable sources of hydrogen all use electrolysis."

Sarah looked a little bit relieved, but I guess she just couldn't resist asking, "When the science teacher did his pop thing, would there have been some water formed again?"

Thomas laughed again. "Of course," he said. "In fact, that's the reverse of electrolysis, and it's more or less what a fuel cell does." He continued: "One of the breakthroughs for all kinds of electrolysis came around 2003 or 2004 when BC Hydro became more open to buying power from independent power producers, because this made lots of the independents economically feasible."

Grandpa interjected: "I remember that. Bruce Sampson, a VP at BC Hydro, realized that by getting people to use energy more efficiently they could save enough money on new equipment to be able to give more encouragement to independent power producers. Higher prices, more flexible standards, that sort of thing. They estimate they've saved something like 7 gigawatts of installed capacity in the last twenty years."

"Right," said Thomas. "Anyway, putting this thing up had its difficulties. The blades are about 50 metres long. They couldn't be brought in by road. Everything came by boat, along with two enormous floating cranes to raise the tower, hoist up the nacelle, install the turbine, and stick the blades on. It took almost three weeks!"

"So what does the power cost?"

"That depends on the wind. Strangely, it works best when the wind is moderate. In very high winds it has to be shut down because the blade tips could break the sound barrier and shake everything to pieces. Of course, costs also depend on how many years the capital cost is amortized over. But if the turbine can be on-line at an average output of, say, 4 megawatts for about one-third of the year, the cost is around 6 cents a kilowatt. For British Columbia, that's just on the margin, but in many places it would be a bargain.

"The trick is to find locations fairly near the grid and to have a way to deal with variable winds. The technology is even better now. Turbines are more efficient, capital cost is down to something reasonable, operating limits are better known, and so forth. But the latest 7-megawatt units are probably about as big as they'll go. I think we'll see lots more of these, in addition to the ones that are now off the west coast of Vancouver Island and in the Queen Charlottes."

Thomas was getting carried away. "Wind power really works great for hydrogen refueling. That hydrogen costs only about $3 per kilogram, so the buses are very cheap to run. Given the efficiency improvements you get from fuel cells, these buses are more than competitive with natural gas. And now you can refuel your car there as well."

Sarah said, "Well, Thomas, that's all very well, but I hear birds are getting chopped up by the blades."

"Just a few, Sarah. Mostly seagulls. The big worry was the eagles that gather nearby every year to eat the dead migrating salmon. None of them was ever hit. We think it's because they can hear the sound, and, anyway, they seem to fly high except upriver."

I interjected: "So, why don't we have these big turbines everywhere?"

"We will before long," said Thomas. "First, they have to become ultra-reliable. That means trouble-free operation for at least 45,000 hours, or around five years. That means gearless, and gearless is simply harder to do. The problem with getting anything new to be certified for 45,000 hours is that it has to really show it

THE PROBLEM IS THAT WHILE THERE'S AN AWFUL LOT OF DIRECT SUNSHINE OR TIDAL POWER OR WIND POWER, THERE SIMPLY ISN'T ENOUGH AVAILABLE AT REASONABLE COSTS TO REPLACE ALL THE HYDROCARBONS WE USED TO USE.

can do it, and that takes five years, and by the time five years have gone by, it's obsolete and you have to start all over again!

"But even then these turbines won't be enough to replace oil and gas. Maybe that will happen here, but around the world we can see now there's about to be a huge expansion in the use of nuclear power."

Sarah objected: "I think that's awful. Why do we have to turn to nuclear power?"

Grandpa tried to moderate: "It would be great to think we could depend only on solar energy in one form or another, or on other forms of non-destructive energy like geothermal, to provide all our needs in the long run. The problem is that while there's an awful lot of direct sunshine or tidal power or wind power, there simply isn't enough available at reasonable costs to replace all the hydrocarbons we used to use.

"For sure, the big rise in oil and natural gas prices has encouraged alternatives. Energy efficiency improvements are terrific compared to how wasteful we used to be. But the whole world insists on having a high standard of living — and who can blame people for that! But it still means using a huge amount of energy. Only nuclear power can fill the gap. Old-style nuclear plants were far too expensive, though. Now, nuclear power has become a lot safer, especially with the new, small fusion reactors, which don't require the massive central generating stations of the old days. They don't contain anything that would be useful as a terrorist weapon, and the short-lived radioactivity they produce has largely done away with fears of long-term radioactive waste. In fact, as Geoffrey Ballard and many others were saying way back twenty years ago, the most likely source of hydrogen would come from electrolyzing water using electricity generated through nuclear power. So, getting to the hydrogen age had this other roadblock, because environmentalists who wanted hydrogen power couldn't stand the notion that more nuclear plants were needed to achieve that end."

Sarah wanted to get back to the topic of fuel cells: "Grandpa, who decided to start building hydrogen gas stations? Why would the service station business have spent money on hydrogen equipment before there were enough fuel-cell cars to justify it?"

Grandpa looked grateful for her change of subject. "The fuel supply issue wasn't minor. Most of the early supply of hydrogen had to come from 'reforming' fossil fuels, because the renewables just weren't cheap enough. Reforming is the process of stripping hydrogen from natural gas or oil or whatever. This is where the environmentalists had a concern, because the residue from the reformer is carbon dioxide and a bit of carbon monoxide. And both of those had to go. If the reformers are big enough, you can actually capture the carbon dioxide and do something with it."

"Like what?" asked Sarah.

"Well," said Grandpa, "for starters, you can 'sequester' it. This means hiding it in places where it'll stay for a very long time. One of the neat things we can do is to pump the carbon dioxide down some kinds of coal mines and drive out methane to be captured and used. It's a lot cheaper and makes far less greenhouse gas than mining the coal and using it for power."

Sarah kept up her questioning. "Grandpa, I gather that this is what they do up in the Alberta tar sands. What's that about?"

"That's another long political story," said the old man. "In a nutshell, the Americans wanted desperately to get their hands on the oil from the tar sands, because they had few other secure sources after the revolts in the Middle East ten years ago. But the tar sands were already the largest single-point source of carbon dioxide in the world. The federal government insisted that all new production had to sequester at least 95 percent of the extra carbon dioxide, and this looked like it was going to be very expensive. The Americans were very unhappy, because they saw this as interfering with a continental oil policy. Canadian sovereignty was at stake. Eventually we agreed on an 80 percent compromise, but you aren't the only person in this country who's very upset about the whole thing."

Thomas then asked a question that hadn't come up before. "Grandpa, there's something else I don't understand. I thought that another major problem with the tar sands was that the producers need huge amounts of hydrogen in order to turn the sticky goo into something manageable. Didn't this distort the whole supply structure?"

Grandpa grimaced at the question. "Well, Thomas, that was one of the problems with getting to the 'hydrogen age.' Here's the issue: were we better off using hydrogen to refine the tar sands, so we could access the hydrocarbon and turn it into gas for our cars, or were we better off simply to use that hydrogen in our cars as it is? Ultimately it was big oil that called this tune, and they were happy to ensure that the United States could continue to exist for some time without having to build a new hydrogen infrastructure."

Sarah said, "Thanks, Grandpa. Sorry I got into that! Can we get back to the local gas stations?"

"Sure. The need for reforming raised some interesting choices. For high-temperature fuel cells, the kind that make electricity for the electrical grid, reforming isn't an issue. They use natural gas directly. But the lower-temperature fuel cells, the kind used in vehicles, need pure hydrogen. The question was whether reforming could best be done in a small device in the car itself, so that the vehicle could be fueled normally and make its own hydrogen on board, or whether it could be done better in convenient batches at some stationary location where vehicles could come and take on hydrogen.

"Back in the early 1990s, I used to argue about this with Firoz Rasul, the far-sighted CEO of Ballard Power. I've always been an advocate of refueling with hydrogen made and delivered at a local gas station. Firoz thought the decision should be made by the customer (or maybe an interim customer such as the car manufacturer). Ballard was prepared to move along both paths.

"Firoz would emphasize that hydrogen is very hard to carry. Back then, no one had settled on the answer we have now, which is using compressed organo-metal hydrides. The options were to have it frozen or compressed to really high pressures, such as 10,000 pounds per square inch, which sounded too much like a bomb to many people. And it took a long time to dissipate the notion that hydrogen is dangerous — the '*Hindenburg* syndrome.' We hoped we could find some other fancy technology, such as carbon nanotubes or metal hydrides. All of these had some disadvantage — they were either too heavy, or they had to be a certain shape, or they needed lots of energy to compress, or freeze, or release the hydrogen from the hydride.

"In my arguments with Firoz, I would insist that the alternative to carrying pure hydrogen would have to be to carry a liquid fuel like methanol or even gasoline, accompanied by a small chemical-reforming plant right in the car or bus — which would never allow the capture of carbon dioxide or other emitted gases, as could be done with a stationary reformer. Since a car is only used around 5 to 10 percent of the time, this would be a very expensive chemical plant.

"Firoz would point out that my scheme required someone to build an infrastructure of gas stations before there were enough customers to justify them, whereas the reformer in the car created the infrastructure exactly in step with customers. It didn't help when articles appeared in major newspapers saying that getting all of the United States rigged for hydrogen would cost up to $30 billion. That seemed like a big number, but it was much less than half of what the country spent to attack Iraq in 2003."

"When did the big multinationals get into this?" asked Sarah. I knew she was suspicious of anything multinational.

Grandpa had probably anticipated this question. "In the early part of the century. Shell Hydrogen, for example, invested in technologies adapted to local service stations — natural gas reformers, hydrogen storage, delivery methods, and so on. Shell built gradually, starting with tube trailers carrying hydrogen produced in a central facility, migrating to skid-mounted reformer units capable of servicing perhaps one hundred cars per day, this being twinned or tripled as demand grew. Eventually, the station would get a new unit capable of handling around 400 cars a day, at which point the trailers and the smaller reformers would be moved outward again to the fringes of demand. This made sense because the capital cost would be spread out over time, and the gas companies could see how they'd get a return sooner.

"In early 2004, Shell and DaimlerChrysler demonstrated the technology by installing some units in Washington, D.C., that could service maybe twenty-five cars per day. And they pulled in the other gas giants: Chevron, Texaco, and Exxon. Small companies like H_2Gen sprung up with very innovative answers. Several Japanese firms put up demonstration stations in Tokyo at about that time.

"In the end, that's what happened. It turned out that the cost of installing the infrastructure was very competitive, since the original fuel was delivered by pipelines already in the ground instead of by great big trucks, and because the costs of maintaining gasoline stations to enhanced environmental standards

THE TRUTH IS, IF WE'D BEEN USING HYDROGEN FOR FUEL, AND SOMEONE SUDDENLY SUGGESTED USING GASOLINE INSTEAD, GASOLINE WOULD NEVER GET PAST THE SAFETY FOLKS BECAUSE IT'S WAY MORE DANGEROUS.

had to be taken into account as well. Getting the infrastructure established was never the obstacle to having fuel-cell cars that the pessimists predicted."

I wanted to return to a couple of things that Grandpa hadn't talked about. "Grandpa, you talked about the *Hindenburg* syndrome, and I know that hydrogen is so light that it rises fast. But what about the explosiveness? Also, I still hear people say that because hydrogen is so light when it escapes, it goes right to the top of the atmosphere and becomes a greenhouse gas all on its own. Are these just perceptions or are they real?"

"Well, Nicholas," Grandpa responded, "it turns out that for the *Hindenburg*, I read somewhere that the outside cover was coated in stuff that NASA later used for rocket fuel, and that's what you saw burning on those famous film clips. And most of the people who died did so from jumping, not from getting burned. The truth is, if we'd been using hydrogen for fuel, and someone suddenly suggested using gasoline instead, gasoline would never get past the safety folks because it's way more dangerous.

"And on the greenhouse gas effect, it's interesting that no one raised much of a stink about this when we were already making many thousands of tonnes of hydrogen for industrial uses. It was just when people started to talk about replacing gasoline in cars that some researchers suddenly discovered and then publicized this scare that the hydrogen would rise, combine with oxygen to make water very high up in the atmosphere — higher than where water normally goes through evaporation — and then become a greenhouse gas. It's true — that does happen. But the quantities are very small, and we now control hydrogen leakage to tiny amounts, and the impact is still over-whelmed by the amounts of carbon dioxide we make and the methane that escapes. So it was a catchy story, and it caused some of my friends in the fuel-cell business a bit of panic — but only for about a week."

ELIZABETH AND REBECCA

On the next weekend, Grandpa and I visited my cousin Elizabeth, Sarah's older sister and a high school science teacher. Elizabeth lives and teaches in Yaletown, the informal centre of Vancouver's thriving software and wireless technology community. Yaletown started in refurbished old warehouse spaces at the turn of the twentieth century, but got built out into new high-rises along the shores of False Creek when we were all kids. It's still the best place in town to live, and Grandpa also hangs out down there.

Just for the summer, Rebecca, another cousin of ours, was living with Elizabeth. 'Becca was a post-doc in economics at Simon Fraser University, and so was a teacher as well. I guess that's what she and Elizabeth had in common.

Elizabeth was pleased about some things at her work. "The kids in my classroom all use personal information devices (they call them 'pids'). They plug into a separate fuel-cell pack you carry around on your belt. They're great. A pid is sort of like a handheld version of the old laptop, and then some. They do everything wirelessly. They have cell phones, Internet browsers, file trading, secure meetings, cameras and picture distribution, music, radio and television, and so on. I send my kids stuff during class — like diagrams and notes.

"These things never used to work. Batteries wouldn't last for the whole day, and the cost of getting electricity to each seat was way too much for school boards. Now the small fuel cartridges last eight hours each and sell by the dozen for the cost of a cup of coffee. Those little fuel cells are such a great innovation. Grandpa, I'm wondering if you know how they were developed?"

Grandpa replied, "I know something about it. The problem was that wireless applications just ate batteries for lunch. In spite of the power-saving software, like the old Centrino that Intel put into their built-in wireless chipsets, something better was needed. Besides, some of those jazzy new batteries were unstable — a laptop once blew up on a baggage conveyor belt in San Francisco Airport.

"Around 2003 or so, many start-up companies began coming up with small fuel cells that used methanol, which is basically the alcohol version of natural gas. The big Japanese consumer-electronic companies, like Toshiba, Sanyo, and Matsushita, and chip companies like Intel, all became potential customers. Before long the start-ups decided they'd better work

BREAKTHROUGHS ONLY OCCUR WHEN USERS CAN SEE SOME COMBINATION OF BENEFITS THAT ADD UP TO BEING FIVE TIMES BETTER THAN WHAT THEY PRESENTLY USE.

together, especially when some laptop manufacturers began to buy components and do their own assembly. Soon the whole thing got commodified. The fuel cells were cheaper than batteries, even including methanol cost. But cost alone wasn't everything. Remember the famous five-times principle."

"Which means?" asked Elizabeth, rolling her eyes.

"The idea is that breakthroughs only occur when users can see some combination of benefits that add up to being five times better than what they presently use. It's pretty subjective, of course. But if these methanol fuel cells cost the same as or less than batteries, and if the users get much more convenience, and if the pids are lighter and do more things, and you can take them on airplanes and find fuel at the average corner store, and so on, then they very rapidly take over. I know they're all around, but I haven't seen one up close for quite a while."

"Here," said Elizabeth, "I'll go get one."

She came back carrying a small box about twenty-five centimetres wide, a couple of centimetres thick, and maybe twenty centimetres deep. The side had a few control buttons around a screen that could be written on or take voice commands or dictation. Elizabeth was already wearing the fuel-cell pack.

"Where's the keyboard?" asked Grandpa.

"If you want one, you attach it to a port, and then this

becomes the screen. With voice recognition, you don't need a keyboard most of the time, but you can add one when you want. Handwriting recognition is totally standard, or you can store it as written. There's no CD or DVD anymore, or serial ports that you carry around. No more telephone modems or Ethernet modems. All wireless."

"And the cell phone?"

"All done with a headset using built-in Wi-Fi."

"Where's the fuel cell and the fuel?"

"Well, the fuel-cell pack is tiny now, as you can see — maybe seven by ten centimetres, and thinner than the pid itself. And if you look here," she took the pack off her belt, "methanol cartridges slot into here — always two, so you never run out. Because it's separate from the actual device, the people who make the pids or the cell phones or the portable music players don't worry about fitting batteries, and the devices are smaller and lighter. We can get eight hours on the pid from each cartridge, even if the thing is on all day using wireless. And now that Eveready and Bic and Gillette distribute the cartridges, you can get them anywhere. They sell for about $1.50 a dozen."

"Wow," said Grandpa, "you can really see how the fuel cell meets the five-times principle! A big change from my days. Back then it was the technology developers who tried to drive the industry by pushing innovations. Now change is driven by the people selling applications, who pull the innovations out of the technical people!"

I piped up: "I know an application niche that drove the actual fuel-cell industry. That's the Zamboni!"

Grandpa nodded. "That's actually a bad pun, too. The fuel-cell people found in talking to the architects in ice rinks that because Zambonis ran on propane, to keep from poisoning people with the fumes, the amount of air circulation equipment in any decent-sized ice rink had to be five times larger than otherwise.

Since this stuff all hung from the roof, reducing it by 80 percent could completely change the construction method of building roofs for ice rinks. The Zamboni fuel cells therefore paid for themselves before the rinks were finished because of the lighter construction and the savings on HVAC equipment."

Elizabeth asked, "So why did it take so long to get fuel-cell Zambonis?"

"Well, for two reasons. First, the savings could only come when you built a new rink. Second, it took a while for the fuel-cell folks to realize that they had to turn from being technology oriented to being applications oriented. Instead of proving something could be done and then trying to sell it, they had to start by finding things that were needed and then figure out how to make fuel cells that could do the job. This eventually happened, of course, but back when the Zamboni and the forklift applications were being developed, very few people realized the importance of starting with needs."

I asked, "Is that how you and Chris Reid got started?"

"Exactly. He and I collaborated on the basic technology behind one fuel-cell device, and after that, we sat down to find a really useful application that could be justified economically. We set up Cellex, believe it or not, more than twenty-five years ago.

"We focused on battery-powered forklift trucks. We'd found that in big warehouses the battery didn't last through a shift even when new. The operators had to stop for twenty minutes at least once during the shift, drive over to the charging stations, roll out this one- or two-tonne battery, hook it up to the charger, and roll in another one. For twenty-four-hour operations, there had to be at least three batteries per truck, and a charging station, and the real estate for that, and probably a guy whose only job was to ensure that the charging was being done properly — keeping battery acid off the floor and so on.

"We calculated that a hybrid combining a smallish hydrogen or methanol fuel cell with just the right amount of battery storage for a buffer would be very competitive. This was the first commercial use of fuel cells.

"Of course, it took Cellex several years to get the right fuel cell, but when we did — we got it from a major component supplier — the whole thing took off. That niche strategy turned out to be exactly what happened in the rest of the fuel-cell industry, because people could buy the building blocks from a variety of suppliers. Our vision was to fit the technology to the customer's needs, not the other way around as many others in the early days were prone to do. We just fitted the system to the customer's needs.

"Our competitors tended to get hooked up with particular suppliers. But we avoided that. We believed that applications companies win on marketing, and marketing was what Cellex was about. The surprise was that marketing would turn out to be so very important for the fuel-cell business, compared to technology. I had earlier thought that maybe fuel cells would just be replacements for internal combustion engines, and that the 'industry' would focus on niche cases in which the internal combustion engine would be vulnerable. Instead, whole new ways of doing things won out.

"Another example of the emphasis on applications rather than on technologies was at Chrysalix, that fuel-cell and hydrogen venture fund that Wal van Lierop and I started in, let's see, about 2001. One of our projects was to help commercialize a different type of fuel-cell technology invented at a western U.S. power utility. The engineers had figured out how to swap out parts of the stack — individual cells — while the whole thing kept operating. In other words, you could avoid downtime. 'No downtime' became their sales theme when they found that phone and cellular companies were encountering all kinds of problems with the big lead-acid battery systems used for backup power when the traditional circuits went down. During the big blackout of 2003, the cell phone operators had to send around trucks with portable generators to cell sites to recharge these old batteries. Of course, this was a hit-and-miss affair, and they had a lot of very angry customers who lost service. Replacing batteries with fuel cells in those systems became their first application.

"We persuaded them to take the next step: the sales theme of 'no downtime' became more than just another fuel-cell application. It became the reason for what the company did, the strategic focus of their activity. They learned how to combine the fuel cells with different types of battery, how to get the right direct-current voltages to suit the customers, how to size systems, and so on. After a few years, they abandoned making fuel cells. Instead, they bought parts from others to get the costs down and came to dominate the reliability game."

Rebecca had been paying attention to all this, and since she was doing a paper on commodity sales taxes, she wondered if our grandfather would have some opinions on this subject, as he did on most subjects. "Never let the facts interfere with your opinions," he used to say.

"Grandpa," she said, "I know we have artificially high prices for gasoline now, maybe even more than the Europeans, and we pay extra for natural gas, even though the prices would be high anyway, and there are lots of complaints. How does the government get away with it?"

"That's a long political story," he said. "Back in the 1990s, the federal government promised the Alberta government that they would not levy a carbon tax, which the provincial government was convinced would be the end of the road for its oil and gas business and for its main government revenue. But carbon taxes had to be imposed to meet the targets of Kyotos I, II, and III. Businesses complained that they couldn't pass the carbon tax costs along to their customers and still stay competitive with the United States. Some manufacturing companies extracted higher subsidies, which made the Americans threaten to impose more duties, as they'd done with softwood lumber, which still hasn't recovered.

"Canadians finally came up with a neat answer. First, the federal government agreed to share the proceeds of the carbon taxes with the provinces. Second, they reduced income taxes by an amount equal to the carbon tax proceeds, and passed legislation preventing the carbon taxes from being raised for twenty years. Since this was done over time, people had the opportunity to buy new cars and equipment that used less of the heavily polluting fuels. The oil and gas business didn't mind all that much, because prices were going up anyway as supply declined and the United States took all of the surplus. What emerged was an economy that gave everyone, businesses included, more opportunities to seek the lowest-cost solutions. We all became a lot more efficient, and there were no subsidies the Americans could complain about."

Rebecca said, "I've been told this also improved our average incomes, but I don't see the connection."

"It's the old story: the fewer impediments there are to letting people freely seek their preferred ways of doing things, the more they do those things. It took a long time to get the balances right. What the governments were really doing was turning to a system of full-cost pricing, so that the prices of goods included hidden costs."

I was trying hard to follow all of this, but my math degree wasn't helping me in this instance. So I asked, "What do you mean by full-cost pricing, and what hidden costs are you talking about?"

"Well, here's a simple case. For years we had known that local air pollution caused more emphysema among the elderly, and treating them cost money for the health care system. On top of that, air pollution actually impaired the ability of people in certain areas to learn economic skills, and over time the economy suffered by having to make bigger welfare payments. So cheap energy actually resulted in all kinds of costs that were not reflected in the energy price. Now, with cleaner air, we save money on health care, people live more efficiently, and they have extra money to spend because income taxes have been reduced. Another way to put it is that an apparently more expensive technology actually brought savings across the economy that were not reflected in its higher price.

"The really big struggle was to identify hidden costs and savings in relation to impacts that weren't local and obvious, but global and deep, like climate change. But as that got addressed, our minimal carbon dioxide taxes began to seem like a good investment."

Rebecca changed the subject a little. "Where do the new regulations fit into this? I always thought regulations were to make sure things were done a certain way, mostly for safety and stuff like that."

Grandpa responded, "Some regulations are like traffic rules. We need those. An example is the big fuel-cell demonstration project sponsored by the province early in the century. To get some fuel-cell installations for standalone power actually sited, there had to be new regulations for buildings, hydrogen filling stations, and so on. Getting those regulations right required a heroic effort, which was bolstered by the need for the fuel-cell industry to set standards. Industry worked together and with government to accomplish this new set of regulations, and this co-operation was a terrific by-product of the demos.

"Other kinds of regulation are designed to change how things are done without specifying exactly what method should be used, maybe because the regulators aren't sure what will work best and because they want entrepreneurs to come up with new solutions. We call those 'forward-looking' regulations. Here's an example. In the mid-1980s, a local government organization in the Los Angeles Basin called the South Coast Air Quality Management District, or 'Sqam-d' as we called it, made up of representatives of the local governments in the area, studied the effects on health and on health care costs of local air pollution in the basin. Their findings horrified them. The U.S. government was still in the grip of the car companies who believed that whatever was good for them was good for the country.

"But Sqam-d had been given some powers by the state. Jim Lents, the CEO, and Alan Lloyd, the chief technology officer, drove the process. Sqam-d decided that by certain dates in the future, if a car company was going to sell cars anywhere in the basin, then a certain percentage of them had to meet well-defined anti-smog standards. Some had to have *zero* output of nitrogen oxide, sulphur oxide, particulates, carbon monoxide, and so forth.

"The car companies screamed and yelled that zero could never be achieved. At least not in the time allowed. But Sqam-d was determined to stick by its guns. Some of us saw that only fuel-cell vehicles could solve this problem, as well as help tackle the global

THE PROBLEM IS THAT MUCH OF ANY ECONOMY PROSPERS FROM THE STATUS QUO. BIG CHANGES ARE RARELY EASY, AND SOME PEOPLE INEVITABLY GET HURT

warming crisis. In fact, I believe that if it hadn't been for the determination of the Sqam-d board and key members of its staff, there would be no fuel-cell business at all today. I was so persuaded by their logic and stubbornness that this became a key strategic reason for my venture capital fund investing in Ballard back in 1987. In fact, it was so important for us to understand this that a retired director of Sqam-d, Larry Berg, later became a director of Ballard.

"We made a bet that some car companies, probably not American ones, would find answers. The American car companies, notably GM, came up with pure electric cars, but we knew they would never get the range and performance that customers needed. We cynically wondered if those cars were created to prove that the standards couldn't be met.

"These regulations gained some currency, and eventually they were adopted by several other states. The California Air Resources Board, a state agency, adopted many of them, and despite being fought hard by the car companies, especially the American ones, the rules stayed. And, of course, it was the foreigners like Daimler Benz and Toyota who came through with the fuel-cell answers."

"Did we get those same rules in Canada?" asked Rebecca.

"There wasn't the political will at first. Not until our big demo project showed the province what fuel cells could do for the economy. Then it happened," said Grandpa.

"The problem is that much of any economy prospers from the status quo. Big changes are rarely easy, and some people inevitably get hurt, and some did. But society as a whole is better off. Thank goodness the young people of your generation understand and support the notion that sustainable development practices are not only healthy but necessary if we're to survive."

SOPHIE

My older sister Sophie and her husband have just moved into a new four-storey condo near the top of the south-facing hill in White Rock, a bedroom community south of Vancouver, near the U.S. border. From their balcony, they look right across the border into Semiahmoo, where Grandpa used to golf with some of his pals. They are very proud of participating in a small electric grid co-op served by a multi-home fuel-cell generator — especially since they don't get any bills from BC Hydro! I visited her for dinner recently with my parents, my sister Natasha, and Grandpa Mike.

"Tell me how this works, Soph," I asked when she took us over to see the installation.

"Okay. First, it runs on natural gas, which makes it easy to get fuel because the gas pipeline has been in the neighbourhood for years. Second, it ties in to the other source of electricity and heat we have, which is solar. During the day, we can get electricity directly from the sun. It's much sunnier out here in Surrey than where you live in North Vancouver. We have photovoltaic panels covering the roof. They are the new ones, so they convert almost 14 percent of the sun's energy into electricity. But the roof area isn't very large, and the sun doesn't shine all the time. We have some deep-discharge batteries to store some of that power. It's all automatic. Most of the day the fuel cell runs at a minimum level, turning out heat and electricity, and at peak periods, like dinnertime, it runs at maximum. If the batteries are charged and the combination makes too much electricity, the co-op sells the rest back to BC Hydro."

Grandpa asked whether they used direct or alternating current in the house. He explained these to me: "Both fuel cells and photovoltaics turn out direct current, not the alternating current we're used to. Many electrical appliances we use actually are just as happy with direct as with alternating current. In fact, some of them need direct current and require an internal power supply that converts normal alternating current to direct. Computers, television sets, radios, stereos — all convert to direct. Others, like incandescent lights or toasters or stoves, can use either — they don't care. Fluorescent lights, including the energy-saving types, need alternating current. Appliances with motors in them prefer alternating current because DC motors are bulkier and harder to control. Vehicles all use direct current, and have for years, because that's what batteries require."

Sophie knew the answer to Grandpa's question. "We run on direct current, so we have slightly different plugs. The fridge is solid state now, so it doesn't have a compressor and has a much smaller motor. Our appliances are a bit cheaper to run than before, but of course we have to buy specific types, and we still need some special converters to get the voltage step down from the 48-volt co-op system. It was a definite hurdle to get these systems installed."

"How about heating?" asked Grandpa.

My sister was really into the technical details. She explained, "This is a solid-oxide fuel-cell system, which uses natural gas without reforming it. It runs at 400 degrees, and the excess heat goes through a heat exchanger, which heats water centrally. This gets piped to all members for space-heating radiators and for washing and laundry. We can sell the excess electricity to BC Hydro. The co-op tells me that 85 percent of the energy in the natural gas gets converted into either heat or electricity. We still cook with natural gas, though the price is much higher than it used to be."

"Who bills you?"

"The co-op gives us a monthly bill for all the energy we use, even the natural gas we use for cooking. They decided to amortize the cost of the 50 kilowatt fuel-cell system over ten years, and the solar panels over twenty, so that's a constant figure which everyone pays. Then there's some new software that tracks who uses the power and when they use it. I think last month we paid about $275 everything considered. That's for all our electricity, all our hot water and house heating, and all our cooking."

On the way home, Natasha asked Grandpa if he'd been able to foresee all these changes and predict the ways in which fuel cells would become commercial.

He said, "We made it our business to look ahead at Chrysalix, our fuel-cell fund. We worked hard at analyzing how various markets and technologies would deal with the chicken-and-egg problems of becoming commercial. We used to call the exercise 'making roadmaps.' One of our investment managers, Christine Bergeron, would put together our roadmaps, and then, based on world events, update our predictions every two months. By 2003 or 2004 we were getting pretty good at forecasting. I'll get one of those forecasts for you. They were classics."

"You know, Grandpa," I asked, "after listening to you for the last few months it seems to me that you were quite involved in all this. Did you ever write it down?"

Grandpa nodded. "Well, Nicholas, I did once. Some of it, anyway. It was a chapter in a book about fueling the future, way back in, let me think, about 2003."

"Can you get me a copy?" I asked.

"Let me see if I can find one."[1]

TIMELINES FOR FUEL-CELL MARKET ENTRY (CHRYSALIX PERSPECTIVE)[2]

	Micro Power for Electronics (< 50 W)	Compact Portable (50 W–3 kW)	Stationary Power Plants (> 3 kW)	Mobile (10 kW–250 kW)
Potential Markets	Laptop chargers, docking stations, cell phones, digital cameras, "convergence devices"	Backup power, APUs for RVs and trucks, rechargers, forklifts, cleaning machines, scooters, etc.	Backup power, remote power systems, quality power, distributed generation, home cogeneration	Buses, military and fleet vehicles, cars
Status of Beta Testing and Demos	Extensive betas; first commercial launches announced	First commercial products available (e.g., Ballard's AirGen); many beta units in the field	More than 300 units in operation; many new energy parks or neighbourhood experiences announced	Many beta units in the field (30 new buses in Europe, GM/Hydrogenics military vehicle; Toyota, DC, and Honda fleet vehicles)
Early Market Opportunities Driven/Held Back by:	2004–2007: Drive for wireless applications and battery problems in commercial products already available	2003–2005: Economics, environmental, and grid/power-cord independence issues	2005–2009: Push back by incumbent players (utilities less willing to experiment than a few years ago)	2006–2010: Satisfaction with incumbent technology and non-availability of H_2 expected to result in less than 5 percent of new cars to be FC cars by 2010
Mass Rollout	2007–2010	2005–2009	2009–2012	2010–2015
Major Commercialization Issues	• Size reduction • Thermal management • Cost reduction	• Startup reliability • Environmental operating conditions • Cost reduction	• Cost reduction • Lifetime and reliability • Grid connection	• Cost reduction • H_2 availability • Reliability • Regulatory issues

AFTERWORD

THE CABIN IS CALLED Four Winds because of its complete exposure to the elements. It sits on a two-hectare, wind-scarred island that juts out into the waters of Georgian Bay, at the northeastern edge of Lake Huron. There is no hydro here, so the power comes from a hodgepodge arrangement: a fireplace for heat, solar power to pump the water, propane for the fridge, and kerosene lanterns for light. As we write, the northwest wind howls through the windows and the white-capped waves crash against the shoals. This is a good place to think about the most elemental force on Earth: energy.

The Ojibway hunters who lived here for centuries believed in a moving spirit, Gitchi Manitou, the great animating presence of the world. According to them, the northwest wind was gifted by him, blowing down as it does from Manitoulin Island, the world's largest island in a freshwater lake. This is all a reminder of just how intimate a force energy is in our lives. Energy is not just about billion-dollar megaprojects and the cars we drive, although these are crucial; it's also about the primary elements around us — the water, the sun, and the wind. These elements have fueled most of human history and continue to shape how we experience the world around us. Energy is one of the animating sprits of our lives, and at Four Winds you can feel the raw and wild forces of a world not harnessed by machines. It reminds us that even now, after all our technological development, we still have many choices available about how we want to build and shape our future. It is not too late.

In fact, this little part of Georgian Bay reflects the kinds of energy choices we now have. The owners of Four Winds have chosen not to be connected to the mainland power grid, but a little way to the north a family has made a different choice. They have dragged hydro lines under water and connected them to their house in order to power all the usual amenities: dish washers, clothes dryers, air conditioners. To the south another family has made yet a different choice; they have erected wind turbines — you can see two of them from Four Winds — to harness the strong and relatively constant winds that press in from Lake Huron. Down the bay another family relies on solar power, while still others function with oil furnaces. Each family has a set of values and needs that justifies their choice. How do you judge which choice is right?

The question is harder to answer than it sounds. Trying to pass judgment on the energy decisions someone else makes can be a sobering exercise in hypocrisy. Think of Four Winds. While there is no mainstream power supply, the only way to get here is to drive a boat. The boat is powered by gasoline. So you ask yourself, am I willing to give up the boat? And to get to the boat you have to take a car from the city. The car pollutes. Are you ready to give that up too? What about the petroleum-based fertilizer needed to grow the feed for the cattle that are used to make the beef that becomes the hamburger you eat as you drive on the tar-based highway up to the cabin? Will you give all that up too? The questions go on and on. These are all hard choices, especially since the energy system we use has made our lives better in so many ways, giving us longer lifespans, increased mobility, and higher standards of living. Do we really want to give up these triumphs for some hazy concept of global warming and pollution? Do we have to?

As we try to answer these questions, it's easy to become overwhelmed by the paradoxes and complexities of our energy system. When that happens, it's natural to back away and begin to rationalize our choices. With such a big, global problem, we think, what difference will one more car make, or one more air conditioner? If we're hypocrites, well then, so is everyone else. The immediate benefits of the energy we use are so evident that it's hard to muster the initiative to attempt to do things differently. What's so wrong with trying to find a bit of paradise like Four Winds, forgetting about changing the world, and simply getting on with the business of making our own lives better?

The logic behind this line of thinking leads to a phenomenon called "the tragedy of the commons," where the interests of the individual undermine the interests of society. First coined by the mathematician William Forster Lloyd in 1833, the idea of the tragedy of the commons was made popular when Garrett Hardin incorporated it into his groundbreaking 1968 essay of the same name, published in *Science* magazine.[1] Hardin postulated that any commonly held resource, such as an open pasture used by herdsmen, would eventually be destroyed because self-interest would triumph over the public good. This was not a matter of malice or greed, but simply one of rationality. "As a rational being, each herdsman seeks to maximize his gain," Hardin wrote. "Explicitly or implicitly, more or less consciously, he asks, 'What is the utility *to me* of adding one more animal to my herd?'" And sure enough, all the benefit of adding an animal goes to the herdsman himself. If the herdsman sells the new animal, for example, he will reap all the financial benefit himself. Meanwhile, the negative consequences of adding another animal, such as overgrazing, are shared by all the users of the common pasture. The herdsman rationally concludes that he has more to gain personally from growing his herd than he has to lose. So he keeps growing it. Logically, every other herdsman on the commons does the same thing. Soon the commons itself can no longer sustain the grazing. This then is the tragedy. As Hardin writes, "Each man is locked into a system that compels him to increase his herd without limit — in a world that is limited. Ruin is the destination toward which all men rush, each pursuing his own best interest in a society that believes in the freedom of the commons." In other words, when there is freedom in a commons, people are rewarded for destructive behaviour.

The concept of the tragedy of the commons is highly relevant to our energy discussion. The benefits of many of our private goods — owning a car, or using electricity from coal to light our homes — accrue almost exclusively to the individual users. Yet many of the costs of these goods — such as the pollution and the associated health and environmental costs — are shared by all of society. In other words, while we keep all the benefits of driving our own cars, we share with everyone else the burden of the ozone and carbon monoxide we create. Most of us do not conclude that we ought to forgo our personal advantages when the costs are shared by society at large — especially when almost no one else is willing to forgo the same advantages. So we act out of self-interest. We spend, we consume, and we grow.

Acting out of self-interest is a rational choice and so, from the individual's perspective, it's completely understandable. From a societal perspective, however, it's potentially ruinous, and on a global scale it's even worse. For instance, governments may ask — as the Bush administration in the United States has done in its effort to avoid signing the Kyoto Protocol — why their country should incur costs to reduce their greenhouse gas emissions when other countries are not required to do so. The leadership asks, "Why should we take on the financial hardship of the commons (or the world) by reducing our goods, when others are constantly increasing theirs and reaping the benefits of doing so?" The relentless logic of this position is exactly how the status quo is maintained.

The most common argument against the tragedy of the commons thesis suggests that the rational self-interest of individuals will eventually lead them to impose regulations to protect all of their private property from general harm. That may be true in theory, but it does not seem to be happening when it comes to our current energy situation. The glitch in our economic system that the tragedy of the commons describes remains at the heart of why we continue to make short-sighted, often disastrous decisions about energy. For example, even though there is widespread scientific consensus that global warming has caused our biological systems to reach their breaking points — what Hardin called the "day of reckoning" — very little is being done about it. In almost every place in the world, our "commons" is experiencing the affects of overuse, including right here on Georgian Bay. In the past decade, the water level around the bay has dropped by more than a metre. Most experts agree that this drop goes beyond the seasonal rise and fall of the Great Lakes, that it points to a more fundamental climatic change, partly a result of global warming. The bad news is that some of the water is not coming back. Lake Huron is mostly a glacial lake, meaning that some of it is not replenished by other sources of water; once the water is gone, it's gone. So the negative impact of the energy world touches here too. But well beyond the shores of Georgian Bay, the effects of the energy we use today are being felt in unanticipated ways that may well be leading us to the tragedy of the commons. From increasing desertification in arid countries to the diminishing of the polar ice caps, energy use is altering the world in a way the engineers of all our successes never imagined.

This is why ingenuity is so important a concept and why it needs to be understood in its broadest sense. Ingenuity is not just about technologies and inventions, although it includes these. Ingenuity is also about the very social structures we live within and the culture we choose to create. Thomas Homer-Dixon labels these technical ingenuity and social ingenuity, and too often we focus on the former and dismiss the latter. That is a terrible mistake. Without social ingenuity we risk succumbing to the tragedy of the commons, our personal interests pitted irreconcilably against the broader interests of society. This is why social ingenuity concepts like carbon credits, green energy taxes, full cost accounting, and the triple bottom line are every bit as important as technological ingenuities like fuel cells and wind power. Fueling the future is as much about social systems and a sense of the public good as it is about supplying new technologies to provide more energy.

But it is a mistake to think that ingenuity itself offers easy or clear solutions. It doesn't. The battle over energy is also a battle of conflicting ideas, as we've seen throughout this book. While Geoffrey Ballard proposes that nuclear energy should be used to make hydrogen in a radical switch away from fossil fuels, Len Bolger and Eddy Isaacs say that a more integrated energy strategy is needed, one where fossil fuels like coal can be burned cleanly to make hydrogen. Peter Fairley talks about investing in new designs for wind turbines, while Hunter Lovins and Wyatt King argue that energy efficiency should be central to our overall energy policies. If the experts don't agree on the way forward, how are we, the public, supposed to sort it all out?

The fact is, healthy disagreement ought to be grounds for optimism, not despair. It points to the possibility that we are not faced with a narrow, grim choice — either pick a painful, belt-tightening way forward or maintain the status quo. As the contributors in this book point out, a deeper understanding of the many energy options we have and the wide array of ingenuity available reveals that our future is actually dynamic and wide open, and waits to be created.

The first steps in this creation are to know what's on the menu and then to make intelligent, informed choices that can, as best as possible, anticipate the consequences for the next generation. It's important to think across generations because every choice we make today is a choice we are making on behalf of other people in the future. This is not just a moral argument, but an economic one. The capital cost of building a nuclear plant or a coal plant or a hydroelectric project or a natural gas pipeline ties up billions of dollars in investment and so commits the next generation of taxpayers and voters to a certain kind of future. The nuclear plants in Ontario are a notable example of this. As Homer-Dixon points out, Ontario will be saddled with debt for decades because of decisions made in the 1970s to use nuclear energy. The massive payments continue even though the nuclear power plants are operating well short of full capacity and suffer from huge operational problems and expensive maintenance issues. Like it or not, Ontario will have nuclear energy for another generation and all the expense that goes along with that choice.

Even the proactive decisions we make now will not have an impact for a generation or more. A good example of this is the switch to hydrogen automobiles — now supported by President Bush to the tune of US$1.2 billion. If, as companies like General Motors predict, it takes ten to twenty years to make fuel-cell cars competitive, hundreds of millions of new internal combustion engine cars will get onto the road during that period of time. Each one of those cars will last about seven to ten years, and each one will belch out about five to six tonnes of carbon dioxide per year. If current trends continue, those cars (and SUVs) will be, on average, less fuel efficient than the cars produced in the 1980s. Do the math: it will take almost twenty to thirty years before internal combustion engine cars are off the road. That's another generation whose energy future is already determined. In the energy world, decisions are costly and the sins of the parents are inevitably visited on the children. With biological systems around the world so fragile and overtaxed, we feel the consequences of those sins more profoundly than ever before. That's why we have to act faster, and we have to think two and three generations ahead.

The battle over energy has many fronts. In the oil fields of the Middle East, in the distant provinces of China, and on the gallows of Nigeria, the military front is never quiet. On the business front, companies vie to control the last great fossil fuel reserves and the latest technologies. Energy is still, to use Daniel Yergin's phrase, the ultimate prize, and it will continue to drive our economies and shape our societies perhaps more than will any other sector. And the fight continues on the home front as well. Every time we turn on a light, or renovate our home, or buy a new car, we are making a decision that has implications and repercussions. The energy battle pits our personal responses against our collective responses, and the only way to win this battle is to understand the goal: to reach a sustainable energy future in which the benefits of plentiful energy supply can be provided to all of humanity, without taxing the environment, our health, or future generations. Is this a pipe dream? Not if we learn how to tap our potentially inexhaustible reserves of human ingenuity.

In his 1947 memoir, *War as I Knew It*, General George Patton left the following lesson for future leaders going into battle. "Never tell people *how* to do things," he wrote. "Tell them *what* to do and they will surprise you with their ingenuity." We know what we have to do in order to build a sustainable future. How we'll do it remains one of the great surprises of the new century.

Andrew Heintzman and Evan Solomon
Point au Baril, Georgian Bay, Ontario
July 27, 2003

ABOUT THE CONTRIBUTORS

DR. GEOFFREY BALLARD, Chair of General Hydrogen Corporation and Chairman Emeritus of the Canadian Hydrogen Association, is acknowledged worldwide as the father of the modern fuel-cell industry. He has received numerous awards for his achievements and is an invited guest of honour and keynote speaker to world business and hydrogen events. He was named Business Leader of the Year by *Scientific American* magazine in November 2002, and he was honoured for Energy Innovation by *Discover* magazine in July 2002. In December 2001, he was featured as a Master of Modern Technology on CBC Newsworld. He received the World Technology Network Award for Energy in 1999 and for the Environment in 2001. *Time* magazine identified him as a "Hero for the Planet" in 1999.

LEN BOLGER spent thirty-one years at Shell Canada, finishing that career as President, Shell Canada Chemical Company, and then Vice President, Research and Technology. After retiring from Shell, he became co-founder and chairman of AdvaTech Homes Canada, a position he still holds. He is also a fellow of the Canadian Academy of Engineering, a member of the board of management of the Alberta Science and Research Authority, co-chair of the Alberta Energy Research Institute, a director emeritus of the Canadian Institute for Advanced Research, a director of React Energy Corporation, and a director of Alternative Fuel Systems.

DR. DAVID B. BROOKS is a natural resources economist who has recently retired after fourteen years with the International Development Research Centre, a Canadian crown corporation that supports research on international development proposed and carried out by people in developing countries. He is now Director of Research for Friends of the Earth – Canada (part-time). Among past positions, he was founding director of the Canada Office of Energy Conservation and director of the Ottawa Office of Energy Probe. Among his books are *Zero Energy Growth for Canada*, *Watershed: The Role of Fresh Water in the Israeli-Palestinian Conflict*, and *Water: Local-Level Management*. He has been elected to the International Water Academy, based in Oslo, Norway.

MICHAEL J. BROWN is an entrepreneur focusing on emerging alternative energy technology companies. He was co-founder and until March 1999 the president of Ventures West Capital Ltd., which became Canada's largest private source of capital for early-stage technology companies. He is a co-founder and the chairman of Chrysalix Energy, a multi-corporate limited-partnership fuel-cell venture fund. He has been a director of eight public and many private companies, including Ballard Power. Currently, he is a director of Fuel Cells Canada and Sustainable Development Technologies Canada. In 2000 he was awarded the Cecil Green Award for Entrepreneurialism by the Science Council of B.C. He was educated at UBC and at Oxford (as a Rhodes Scholar).

PETER FAIRLEY is a freelance journalist based in Victoria, British Columbia. His writings on renewable energy, nanotechnology, and sustainable transportation appear often in *Technology Review* magazine, and his byline can also be found in *The Sunday Times of London*, *IEEE Spectrum*, *AlbertaViews*, and *Canadian Business*. He also teaches environmental journalism at the University of Victoria and serves as vice president and membership chair for the Society of Environmental Journalists. He holds a master's degree in journalism from New York University's Science and Environmental Reporting Program and a bachelor of science degree in molecular biology from McGill University.

DR. AVI FRIEDMAN is a professor of architecture at McGill University. He has designed several housing prototypes which were then constructed in the private sector. The Grow Home, a narrow-front rowhouse, was first presented in 1990 and has since been built in numerous communities across North America. The Next Home, a flexible and affordable housing type, has also been incorporated into the design of communities. His work has been covered on television shows such as *Good Morning America*, *Dream Builders*, and *How Buildings Learn* and in major magazines and newspapers. He has received numerous awards, including the Association of Collegiate Schools of Architecture (ACSA) Collaborative Practice Award, the ACSA Creative Achievement Award, and the United Nations World Habitat Award.

SUSAN HOLTZ is a senior consultant who specializes in energy, environment, and sustainable development. She has been a principal in projects on long-term energy planning, electricity rates, municipal energy management, and greenhouse gas reduction. She also served on government hearing panels on electricity regulation in Nova Scotia and on hydrocarbon exploration on Georges Bank. She has been appointed to many advisory bodies and boards, including the Auditor General of Canada's Panel of Senior Advisors and the External Advisory Panel for Canada's Commissioner of Environment and Sustainable Development, and was the founding vice chair of both the Canadian and Nova Scotia Round Tables on the Environment and the Economy.

THOMAS HOMER-DIXON is Director of the Centre for the Study of Peace and Conflict and Associate Professor in the Department of Political Science at the University of Toronto. His work is highly interdisciplinary, drawing on political science, economics, environmental studies, geography, cognitive science, social psychology, and complex systems theory. His writings have appeared in leading scholarly journals, popular magazines, and newspapers. His books include *The Ingenuity Gap*, which won the 2001 Governor General's Non-fiction Award, and *Environment, Scarcity, and Violence*. He has been invited to speak about his research at many universities, including Yale, Harvard, Princeton, Cornell, Oxford, and Cambridge, and at the World Economic Forum in Davos, Switzerland, and the Council on Foreign Relations in New York.

DR. EDDY ISAACS is the managing director of the Alberta Energy Research Institute. Previously, he served for more than twenty years with the Alberta Research Council, where he was responsible for the programs in heavy oil and oil sands. He currently serves on the boards of the Petroleum Technology Alliance of Canada, Canadian Oil Sands Network for Research and Development, and the IEA Weyburn Monitoring Project, and is the co-chair of the Technology Working Group of the National Air Issues Coordinating Committee on Climate Change.

The Alberta Energy Research Institute (AERI) was formed on August 1, 2000, as successor to the Alberta Oil Sands Technology and Research Authority (AOSTRA) but with a broader mandate. AERI promotes energy research and technology evaluation and transfer in areas that include conventional and unconventional oil and gas, coal, carbon management, improving energy efficiency, and renewable energy as part of a cleaner energy strategy for Alberta.

WYATT KING is an expert in on-line business and international program management. As an early employee at Amazon.com, he managed multi-national, cross-disciplinary teams on projects ranging from opening new customer service facilities in Germany to launching new product lines in France. Educated at Williams College and the University of Washington, he is currently a graduate student at the Fletcher School.

GORDON LAIRD is a journalist, photographer, and author and has worked for *Saturday Night*, *Canadian Geographic*, the *Far Eastern Economic Review*, the *Globe and Mail*, *This* magazine, and *Mother Jones*. He has won two gold medals from Canada's National Magazine Awards, including top honours for investigative reporting in 2001. His specialty is documentary reportage, including photo-text packages and special investigative projects. He has covered stories on the environment, politics, and culture from a wide range of locales, including the high Arctic, Tibet, Mongolia, and China. His books include *Power: Journeys Across an Energy Nation*, a documentary survey of climate change, energy, and the perils of the industrial age. His Web site address is www.gordonlaird.com.

L. HUNTER LOVINS is the president of the Natural Capitalism Institute and is former president of the Rocky Mountain Institute. The co-author of eight books, including *Energy and War: Breaking the Nuclear Link* and *Brittle Power: Energy Strategy for National Security*, she has appeared on numerous television shows, including *60 Minutes* and dozens of news programs. She has addressed such audiences as the U.S. Congress, the World Economic Forum in Davos, Switzerland, the World's Fair Energy Symposium, and the State of the World Forum. As a consultant, she has briefed senior management at such groups as Mitsubishi, Bank of America, and Royal Dutch/Shell Group. Her public sector clients have included the U.S. Defense Civil Preparedness Agency, the Solar Energy Institute, and the U.S. Environmental Protection Institute.

ALLISON MACFARLANE is Associate Professor of International Affairs and Earth & Atmospheric Science at the Georgia Institute of Technology. Her research focuses on the issues surrounding the management and disposal of high-level nuclear waste and fissile materials. Previously she was a Senior Research Associate at the Massachusetts Institute of Technology's Security Studies Program. From 1998 to 2000 she was a Social Science Research Council–MacArthur Foundation fellow in International Peace and Security at the Belfer Center for Science and International Affairs at Harvard University. She has also served on a National Academy of Sciences panel on the spent fuel standard and excess weapons plutonium disposition. She received her Ph.D. in geology from MIT in 1992.

JEREMY RIFKIN is the author of many best-selling books, including *The Hydrogen Economy: The Creation of the World Wide Energy Web and the Redistribution of Power on Earth*. He serves as an advisor to Romano Prodi, the president of the European Commission, and provided the strategic white paper that led to the European Union's adoption of a new energy initiative to become the first fully integrated hydrogen superpower in the twenty-first century. He is also the president of the Foundation on Economic Trends in Washington, D.C. Since 1994 he has been a fellow at the Wharton School's Executive Education program, where he lectures to senior corporate management from around the world on new trends in science and technology and their impacts on the global economy, society, and the environment.

JANE THOMSON is a master's candidate in the Faculty of Environmental Studies (FES) at York University in Toronto. Her area of concentration is "Business as an Essential Partner in Sustainable Development," and she is completing the Diploma in Business and Sustainability offered through FES together with the Erivan K. Haub Program in Business and Sustainability at the Schulich School of Business. She earned a bachelor of science degree in Mechanical Engineering from the University of Manitoba, after which she worked in the consulting industry for five years in Canada, Ghana, Nigeria, Uganda, and Kenya before returning to university to pursue graduate studies.

DR. DAVID WHEELER is the founding director of the York Institute for Research and Innovation in Sustainability at York University, in Toronto, and Erivan K Haub Professor in Business and Sustainability at the Schulich School of Business. The co-author of *The Stakeholder Corporation*, he has been published in a wide variety of journals, books, and parliamentary inquiries and has done numerous television and radio broadcasts. He was Executive Director of Environmental and Social Policy at The Body Shop International for seven years. He is an advisor to the governments of Canada and the U.K., has been a frequent consultant to United Nations and other development agencies working in water and sanitation programs in less developed countries, and has supervised development projects in Africa and Latin America.

KEN WIWA is an author, journalist, broadcaster, and social entrepreneur. A weekly columnist for the *Globe and Mail*, he writes on cultural and political issues. He has presented programs for the BBC and documentaries for Channel 4, CBC, and BBC Radio 4. His first book, *In the Shadow of a Saint*, won the 2002 Hurston-Wright Foundation Non-fiction Award. He divides his time between Canada and Nigeria, where he is the managing director of Saros International, a company that he is restructuring to accommodate and build virtual networks to provide infrastructure and capacity for agricultural, health, cultural, and educational initiatives. He is a Saul Rae Fellow at the Munk Centre for International Studies at the University of Toronto and a senior resident of Massey College.

ABOUT THE EDITORS

ANDREW HEINTZMAN is president of Investeco Capital Corp., a company that provides capital and expertise to environmental businesses. He is also currently a director of The Sustainability Network, an organization that develops capacity building for Canadian environmental not-for-profit organizations. He was the co-founder and publisher of *Shift Magazine* and founder of consulting company d~Code.

EVAN SOLOMON is the host of CBC Newsworld's *Hot Type* and co-host of the current affairs show *CBC News: Sunday*. He co-founded and was the editor-in-chief of *Shift Magazine* and is the author of the novel *Crossing the Distance*. He has hosted many award-winning television series about the impact of technology on society, including CBC's *FutureWorld* and *ChangeMakers* and PBS's *Masters of Technology*. He contributes regularly to newspapers and magazines around the country and sits on McGill University's Arts Advisory Board.

NOTES AND ACKNOWLEDGEMENTS

Introduction
OIL, LIPSTICK, AND FINDING SOLUTIONS FOR THE FUTURE

1. *Agence France-Presse*, June 13, 2003.
2. *Globe and Mail*, June 23, 2003.
3. Ibid.
4. Daniel Yergin, *The Prize: The Epic Quest for Oil, Money and Power* (Free Press, 1993), 337.
5. American Petroleum Institute, http://api-ec.api.org/frontpage.cfm.
6. The British Library, http://www.bl.uk/whatson/exhibitions/evelynnotes.html.
7. Cardiff University, http://www.cf.ac.uk/encap/skilton/nonfic/evelyn01.html.
8. http://www.grida.no/climate/ipcc_tar/wg1/index.htm.
9. Thomas Homer-Dixon, *The Ingenuity Gap* (Alfred A. Knopf, 2000), 8.

Chapter 1
BRINGING INGENUITY TO ENERGY

1. For a complete discussion, see Thomas Homer-Dixon, *The Ingenuity Gap: Can We Solve the Problems of the Future?* (Toronto: Knopf Canada, 2000), and Thomas Homer-Dixon, "The Ingenuity Gap: Can Poor Countries Adapt to Resource Scarcity," *Population and Development Review* 21, no. 3 (1995): 587–612.
2. Quoted in Gordon Laird, *Power: Journeys Across an Energy Nation* (Toronto: Penguin Viking, 2002), 167.
3. Sheela Basrur, "Air Pollution Burden of Illness in Toronto," City of Toronto (Toronto: Toronto Public Health, May 2000): i–ii.
4. Henry David Venema and Stephan Barg, "The Full Costs of Thermal Power Production in Eastern Canada" (Winnipeg: International Institute for Sustainable Development, July 22, 2003).
5. This research is reviewed in Martin Seligman, *Authentic Happiness: Using the New Positive Psychology to Realize Your Potential for Lasting Fulfillment* (New York: Free Press, 2002), 51–55.

6. Laird, *Power*, 144. The phrases quoted in the last sentence are from Laird's interview with Bryan Young, general manager of the Toronto Renewable Energy Coalition.

Chapter 2
AT THE FRONTIER OF ENERGY

1. Carol Howes, "2001: An Ice Odyssey," *Financial Post (National Post)*, September 2000, sec. D: D1, D4.
2. Mackenzie Gas Project, Preliminary Information Package, April 2003, vol. 1, Project Description.
3. L. F. Ivanhoe, "Canada's Future Oil Production: Projected 2000–2020," M. King Hubbert Center, For Petroleum Supply Studies, Colorado School of Mines, (Hubbert Center Newsletter #2002/2).
4. Robert G. Bromley, Project Director, "Progress Report: What Is Sustainable Community Energy Project," *Ecology North*, Yellowknife, NT, 17 October 2002.
5. Ibid.
6. United Nations Environment Programme (UNEP), June 2001.
7. CNN, "Shell to Ink PetroChina Pipeline Deal," June 28, 2002, http://www.cnn.com/2002/BUSINESS/asia/06/28/hk.petrochina/index.html.
8. Geoffrey A. Fowler, "Trouble in Toy Town," *Far Eastern Economic Review*, February 6, 2003.
9. Amnesty International, "China," 2003, http://web.amnesty.org/report2003/chn-summary-eng, and Amnesty International, "China: Torture — A Growing Scourge," February 2001.
10. http://www.shell.com.cn/english/wep/schedule.html.
11. "China's Big Bet on Gas," *Businessweek*, April 29, 2002.
12. BBC, "China's 'War on Terror'," September 10, 2002, http://news.bbc.co.uk/1/hi/not_in_website/syndication/monitoring/media_reports/2241025.stm, and BBC, "US Diplomat Visits China's Muslim Area," December 18, 2002, http://news.bbc.co.uk/1/hi/world/asia-pacific/2586655.stm.
13. Mark O'Neill, "Xinjiang Separatists Called Top Terror Threat," *South China Morning Post*, November 4, 2002; "Uyghur Leader Denies Terror Charges," *Radio Free Asia*, January 29, 2003; and Matthew Forney, "One Nation Divided," *Time Magazine*, March 25, 2002.
14. Stephanie Mann, "China's Report on Xinjiang Region Questioned," *Voice of America*, May 29, 2003.
15. Matthew Forney, "One Nation Divided," *Time Magazine*, March 25, 2002.
16. StratFor, "China: Separatist Crackdown Could Taint Foreign Companies," May 29, 2002.
17. AFP, "Anti-American Sentiment Among China's Muslims Growing," March 31, 2003, and Reuters, "China Quake Highlights Ethnic Rift in Xinjiang," March 25, 2003, http://www.uygur.org/wunn03/2003_03_25.htm.

18. Jasper Becker, "Shell Defends Pipeline Decision," *South China Morning Post*, March 25, 2002, and Ray Heath, "BP, Shell Hope To Fuel China's Growth," *Evening Standard*, London, January 14, 2003.

19. Claudia Cattaneo and Tony Seskus, "Drop In Oil Exploration Spurs New Supply Fears," *National Post*, April 19, 2003.

Chapter 3
SHAPING AN INTEGRATED ENERGY FUTURE

1. Fridtjof Unander and Carman Difiglio, "Energy and Technology Perspectives: Insights from International Energy Agency Modelling" (presentation, IEA World Energy Outlook 2002).

2. Campion Walsh, "EIA [Energy Information Administration]: Canada's Oil Reserves 2nd Only to Saudi Arabia," Dow Jones Newswires, May 2, 2003, Petroleumworld.com, http://www.petroleumworld.com/story1129.htm.

3. Canadian Association of Petroleum Producers, "Industry Facts and Information - Canada," http://www.capp.ca/default.asp?V_DOC_ID=603.

4. Canadian Association of Petroleum Producers, "Action on Energy - Industry Continues to Surpass Gas Flaring Targets," http://www.capp.ca/default.asp?V_DOC_ID=816.

5. Alberta Energy and Utilities Board, *Alberta's Reserves 2000 and Supply/Demand Outlook 2001–2010*, EUB Statistical Series 2001-98 (Calgary, AB: EUB, 2001).

6. The Coal Association of Canada (submission, 59th Conference of Canada's Energy and Mines Ministers, Winnipeg, MB, September 15–17, 2002), 3.

7. This figure was calculated using statistics from the following sources: Alberta Energy and Utilities Board (see note 5); World Energy Council, *WEC Survey of Energy Resources* (WEC, 1995); Richard Meyer et al., *Proceedings of the 15th World Petroleum Congress, Beijing, People's Republic of China, October 12–17, 1997* (John Wiley & Sons, 1998).

8. Canadian Association of Petroleum Producers, "Industry Facts and Information - Oil, Oil Sands and Natural Gas - Oil Sands," http://www.capp.ca/default.asp?V_DOC_ID=688.

9. Syncrude Canada Ltd., http://www.syncrude.com/.

10. National Task Force on Oil Sands Strategies, Alberta Chamber of Resources, *The Oil Sands: A New Energy Vision for Canada* (Edmonton: Alberta Chamber of Resources, 1995), quoted in Bill Almdal, "Oil Sands Update" (presentation, Regional Issues Working Group 2002).

11. Larry Fisher and Louise Gill, *Supply Costs and Economic Potential for the Steam Assisted Gravity Drainage Process*, Canadian Energy Research Institute Study No. 91 (Calgary, AB: CERI, 1999).

12. Alberta Research Council, quoted in Michelle Dacruz, "Better Than SAGD/VAPEX Technology Will Be a Step Forward," *Fort McMurray Today*, June 25, 2003, A3.

Chapter 5
HYDRICITY, THE UNIVERSAL CURRENCY

1. Editorial, "Kyoto — the Worst of Both Worlds," *National Post*, December 19, 2002.
2. J. Byron McCormick, "Statement of General Motors Corporation, Before The Senate Energy and Natural Resources Committee," July 17, 2001.

Chapter 7
THE PERFECT ENERGY: FROM EARTH, WIND, OR FIRE?

1. U.S. Department of Energy, *Press Release*, Washington, D.C., June 25, 2001.
2. For wind turbine statistics, see the American Wind Energy Association (www.awea.org); the Canadian Wind Energy Association (www.canwea.org); and the European Wind Energy Association (www.ewea.org).
3. National Center for Photovoltaics, *Solar Electric Power: The U.S. Photovoltaic Industry Roadmap*, Golden, CO, 2001, http://www.nrel.gov/ncpv/pvplans.html.
4. Mark Z. Jacobson and Gilbert M. Masters, "Exploiting Wind Versus Coal," *Science*, August 24, 2001.
5. Peter Fairley, "Wind Power for Pennies," *Technology Review*, July 2002.
6. This quotation stems from a personal interview with the source as are all following quotations, unless otherwise documented.
7. Peter Fairley, "Solar on the Cheap," *Technology Review*, January 2002.
8. Peter Fairley, "BP Solar Ditches Thin-Film Photovoltaics," *IEEE Spectrum*, January 2002.
9. BP, *Press Release*, New York, November 21, 2002.
10. International Energy Agency, *World Energy Outlook 2000*, Paris, 2000.
11. Eric D. Larson, "The Princeton-Tsinghua Collaboration on Low Emission Energy Technologies and Strategies for China," presented at the CMI Annual Review Hydrogen Meeting, Princeton, January 2002.
12. Howard J. Herzog, "What Future for Carbon Capture and Sequestration?" *Environmental Science and Technology*, April 1, 2001.
13. R. Moberg et al., "The IEA Weyburn CO_2 Monitoring and Storage Project," presented at the Sixth International Conference on Greenhouse Gas Control Technologies, Kyoto, October 2002.
14. J. Reuther et al., *Technical Report Prospects for Early Deployment of Power Plants Employing Carbon Capture*, Morgantown, WV, National Energy Technology Laboratory, April 2002, http://www.netl.doe.gov/publications/others/techrpts/2430-1a.pdf.
15. David Hawkins, Testimony delivered before the U.S. Senate Committee on Environment & Public Works, Washington, D.C., June 12, 2002.
16. Martin I. Hoffert et al., "Advanced Technology Paths to Global Climate Stability: Energy for a Greenhouse Planet," *Science*, November 1, 2002. In the twentieth century, the concentration of heat-trapping carbon dioxide increased from 275

to 370 parts per million, thanks in large measure to power production (power plants produce about a third of Canada's greenhouse gases and about 40 percent in the U.S.). Without dramatic change in the way we use and produce energy, atmospheric carbon dioxide will pass 550 parts per million this century and, if the scientific consensus is correct, warm the Earth by 2°C — warming that may wreak considerable havoc with our climate. Hoffert et al. estimate that simply stabilizing the atmosphere at that level while meeting the rising demands of a developing world will require the installation of 10 to 30 terawatts of non-polluting power generation by 2050. That is a thousand times more than the installed base of wind turbines in service today worldwide.

17. See the American Association for the Advancement of Science's R&D Budget and Policy Program website for detailed analysis of federal budget analyses, http://www.aaas.org/spp/rd/.

18. A. Pape-Salmon et al., *Low-Impact Renewable Energy Policy in Canada: Strengths, Gaps and a Path Forward*, Drayton Valley, Alberta, Pembina Institute for Appropriate Development, February 2003.

19. Mark Z. Jacobson and Gilbert M. Masters, *Science*.

Some of the material presented in this chapter appeared in different form in two stories — "Solar on the Cheap" and "Wind Power for Pennies" — published in *Technology Review* magazine in January and July 2002, respectively. The author would like to thank two editors at *Technology Review*, David Rotman and David Talbot, who both focused his analysis and sharpened his prose.

Chapter 8
BOOM, BUST, AND EFFICIENCY

1. "Clean technologies" refers to appropriately scaled solar, wind, hydro, biomass, and wave/tide energy supply technologies. In a broader sense, it also refers to the wide array of ways to use all forms of energy more efficiently. It does not include nuclear technology, or the centralized renewable technologies.

2. K. Butti and J. Perlin, *A Golden Thread: 2500 Years of Solar Architecture and Technology* (Palo Alto, CA: Cheshire Books, 1980).

3. Energy efficiency means doing the same tasks using less energy. It is not the same as curtailment, which means cutting back activities to use less energy. Conserving energy can be done either by efficiency or by curtailment. Confusion between these concepts is what led Vice President Dick Cheney, architect of the Bush administration's national energy policy, to state that a nation cannot conserve its way to greatness. Expansionists may argue that this is true of curtailment. Only someone profoundly uninformed of economics would make the same argument of efficiency. Doing more with less, paying less, causing less harm to the environment, and thereby enhancing national security not only constitute an effective foundation for a national energy strategy, they are the cornerstone.

4. A. Lovins and H. Lovins, "Mobilizing Energy Solutions," *American Prospect*, February 2002.

5. Early in his administration Reagan ordered the Department of Energy (DOE) to pulp the Agricultural Yearbook, which taught farmers how to use energy more efficiently. The then executive director of the Daughters of the American Revolution (DAR) went to the DOE warehouse at night and loaded a pickup with as many copies as his truck could carry and distributed them to DAR members at their Continental Congress the next day.

6. A. Lovins and H. Lovins, "Mobilizing Energy Solutions," *American Prospect*, February 2002.

7. According to David Roodman, an analyst at the Center for Global Development, the U.S. government subsidizes automobiles at a rate of US$111 billion a year above and beyond what it reaps in auto taxes and fees — an estimate that does not include the environmental, health, and military costs of burning fossil fuels. One recent example of a subsidy is the provision in the Bush economic plan to increase the amount that business owners can deduct for the purchase of an SUV.

   For example: Dodge Durango sticker price US$27,205; Current law Equipment deduction US$25,000; Total tax deduction* US$25,971. Bush economic plan Equipment deduction US$27,205; Total tax deduction US$27,205. This includes the bonus tax write-off enacted by Congress in March 2002 and a deduction for normal depreciation.

   How the write-off works: Hummer H1 sticker price US$106,185; Current law Equipment deduction US$25,000; Total tax deduction* US$60,722. Bush economic plan Equipment deduction US$75,000; Total tax deduction* US$88,722.

   Sources: Detroit News research, IRS, Taxpayers for Common Sense. Reported in a story by Jeff Plungis / Detroit News Washington Bureau, January 20, 2003.

8. Reuters, "Environmentalists Criticize Ford Fuel Efficiency," June 5, 2003, http://www.planetark.org/dailynewsstory.cfm/newsid/21045/story.htm.

9. Transportation Research Board, *Effectiveness and Impact of Corporate Average Fuel Economy (CAFE) Standards* (Washington, D.C.: National Academy Press, 2002), http://www.nap.edu/books/0309076013/html/. This is a conservative analysis.

10. Some analysts doubt that there is any economically recoverable oil in the Arctic National Wildlife Refuge. The U.S. Geological Survey conducted a peer-reviewed assessment that concluded that there is probably *no* oil that would be cost effective to extract at the moderate oil prices discovered in the futures market, forecast by industry and government, and relied upon by the state of Alaska's revenue forecasts. For more details, see the July–August 2001 *Foreign Affairs* article "Fool's Gold in Alaska." The published version and a heavily annotated hypertext version are both available at http://www.rmi.org/sitepages/pid171.php.

11. Simon Beavis, "Volkswagen Unveils World's First 1 Litre Car," May 24, 2002, http://www.wbcsdmobility.org/news/cat_1/news_106/index.asp.

12. Hal Harvey, Bentham Paulos, and Eric Heitz, "California and the Energy Crisis: Diagnosis and Cure," *The Energy Foundation*, March 8, 2001, http://www.ef.org/california/downloads/CA_crisis.pdf.

13. *New York Times*, May 6, 2001. Cheney also said, "Conservation may be a sign of personal virtue, but it is not a sufficient basis for a sound, comprehensive energy policy."

14. R. Smith, "Power Industry Cuts Plans for New Plants, Posting Risks for Post-Recessionary Period," *Wall Street Journal*, January 4, 2002. The article reports data from Energy Insight (Boulder, Colorado), showing that at least 18 percent, or 91 out of a total announced portfolio of 504 billion watts planned for construction, had been cancelled or tabled by the end of 2001. (The 504-billion-watt portfolio included longer-term projects than those summarized at the beginning of this paragraph.) Ms. Smith interprets the reductions as likely to create power shortages; we interpret them as likely to reduce financial losses when demand assumptions prove exaggerated — especially if saving electricity is allowed to compete fairly with producing it.

15. The states most actively supporting energy efficiency programs continue to be primarily in the northeast, the Pacific northwest, and certain parts of the midwest, along with a handful of other states, including California, Florida, and Texas. The average annual spending across all fifty states is US$3.88 per capita. Connecticut ranks first in per-capita program spending at US$19.48. While this overall national trend is encouraging, the research demonstrates that only about one-third of the states account for nearly all (86 percent) of the spending by utilities and states on energy efficiency programs. The report "State Scorecard on Utility and Public Benefits Energy Efficiency Programs: An Update" is available at ACEEE's Web site at http://www.aceee.org/.

16. This and other information about energy efficiency can be found in A. Lovins and H. Lovins, "Mobilizing Energy Solutions."

17. A. Lovins, "Negawatts, 12 Transitions, Eight Improvements and One Distraction" *Energy Policy* 24, no. 4: 331–343. See also World Alliance for Decentralized Energy, "World Survey of Decentralized Energy — 2002/2003," http://www.localpower.org/.

18. Productivity increases when workers can see better what they're doing, breathe cleaner air, hear themselves think, and feel more comfortable. Offices typically pay about one hundred times as much for people as for energy, so 6 to 16 percent higher labour productivity increases profits by about 6 to 16 times as much as eliminating the entire energy bill. See J. J. Romm and W. D. Browning, "Greening the Building and the Bottom Line," *Rocky Mountain Institute*, 1994, http://www.rmi.org/images/other/GDS-GBBL.pdf.

19. Lester R. Brown, Janet Larsen, and Bernie Fischlowitz-Roberts, *The Earth Policy Reader* (New York: W. W. Norton & Company, 2002). Also available on-line at http://www.earth-policy.org/Books/EPR_contents.htm.

20. Reuters, "New EU Law Aims to Double Green Energy by 2010," July 5, 2001, http://www.viridiandesign.com/notes/251-300/00254_europe_doubles_green_power. The strategy was first laid out in the European Commission's "Energy for the Future: Renewable Sources of Energy," section 1.3.1, p. 9, http://europa.eu.int/comm/energy/library/599fi_en.pdf.

21. European Wind Energy Association, "European Wind Energy Capacity Breaks the 20,000 MW Barrier," November 2002, http://www.ewea.org/doc/20gw%20briefing.pdf.

22. Reuters, "German 2002 Wind Power Market Up 22 Pct," January 24, 2003, http://www.planetark.org/dailynewsstory.cfm?newsid=19548.
23. Reuters, "Germany Approves Second Offshore Wind Project," December 19, 2002, http://www.planetark.org/dailynewsstory.cfm?newsid=19129.
24. European Wind Energy Association, "European Wind Energy Capacity Breaks the 20,000 MW Barrier."
25. Reuters, "US Wind Power Growth Waned in 2002," January 27, 2003.
26. Lester R. Brown, "Wind Power Set to Become World's Leading Energy Source," *Earth Policy Institute*, June 25, 2003, http://www.earth-policy.org/Updates/Update24.htm.
27. H. Lovins and W. Link, "Insurmountable Obstacles," (paper presented to the United Nations Regional Roundtable for Europe and North America, Vail, Colorado, 2001).
28. Lester R. Brown, *Eco-Economy: Building an Economy for the Earth* (New York: W. W. Norton & Company, 2001). Also available on-line at http://www.earth-policy.org/Books/index.htm.
29. Shell International, *Exploring the Future: Energy Needs, Choices and Possibilities — Scenarios to 2050* (2001), http://www.shell.com/home/media-en/downloads/scenarios.pdf.
30. M. Mansley, "Long Term Financial Risks to the Carbon Fuel Industry from Climate Change" (London: Delphi Group, 1995).
31. Michelle Nichols, "Solar Power to Challenge Dominance of Fossil Fuels," Reuters, August 9, 2002.
32. Reuters, November 25, 2002.
33. None, the free market will do it . . .
34. C. van Beers and A. de Moor, *Public Subsidies and Policy Failures* (Cheltenham U.K.: Edward Elgar Publishing, 2001). The European Commission took the first step towards possible reform of E.U. energy subsidies by producing an inventory of all forms of state support provided to the fossil fuel, renewable, and nuclear energy sectors. Published in December 2002, the document "Working Paper of the European Commission: Electricity from Renewable Energy Sources and the Internal Electricity Market" says the inventory may determine "whether certain energy sources are . . . given advantages that do not adhere to the objectives of energy policy and combating climate change." In recent years, members of the European Parliament, environmentalists, international bodies such as the OECD, and, increasingly, member state governments have been asking for the support granted to different energy sources to be made more transparent. Furthermore, the E.U. renewable energy directive, adopted in 2002, calls on the commission to propose a harmonized support framework for renewables if a comparison of subsidies identifies any "discrimination" between energy sources. ENDS Environment Daily January 9, 2003.
35. I have sat in meetings with senior energy officials from such European nations as France who have denied that their governments give any subsidies, and when confronted with clear evidence of a variety of subsidies, have refused to reveal the amounts given (for example, to the nuclear programs of Électricité de France, one of the most heavily subsidized electricity programs in the world).

36. R. Heede, "A Preliminary Assessment of Federal Energy Subsidies" (1984), http://www.rmi.org.

37. Ricardo Bayon, *Atlantic Monthly*, January–February 2003, 117.

38. Matthew Wald, "Use of Renewable Energy Took a Big Fall in 2001," *New York Times*, December 6, 2002.

39. Tom Doggett, Reuters, April 2, 2002.

40. Elinor Burkett, "A Mighty Wind," *New York Times*, June 15, 2003, http://www.nytimes.com/2003/06/15/magazine/15WIND.html?ex=1057204576& ei=1&en=c6657f4c00300148.

41. British Wind Energy Association brochure, http://www.bwea.com/pdf/gen.pdf.

42. Howard Geller, *Energy Revolution: Policies for a Sustainable Future* (London: Island Press, 2002).

43. E.U. programs to improve end-use efficiency include the GreenLight Programme, the Standby Initiative, the Motor Challenge Programme, the Luminaire Design Competition, the EuroDEEM database, and initiatives in the appliance, building (e.g., the Thebis database), and industrial fields. See http://energyefficiency.jrc.cec.eu.int/html/readmore.htm.

44. "Energy-Intelligent Europe Initiative," ACEEE, http://www.eceee.org/ eceee_forum/EI-Europe.lasso.

45. European Alliance of Companies for Energy Efficiency in Buildings, address to the Parliamentary Hearings on Intelligent Energy for Europe, http://www.europarl.eu.int/hearings/20020911/itre/euroace.pdf.

46. Sacramento Municipal Utility District, "Facts and Figures," *SMUD*, http://www.smud.org/about/facts/index.html.

47. *Public Utilities Fortnightly*, December 1, 1994.

48. Stephen Beers and Elaine Robbins, "Lights Out: The Case for Energy Conservation," *E Magazine*, January 1998.

49. Ibid.

50. Sacramento Municipal Utility District, "Shade Tree Program," *SMUD*, http://www.smud.org/sacshade/index.html (accessed September 2001).

51. Stephen Beers and Elaine Robbins, "Lights Out: The Case for Energy Conservation."

52. SMUD, http://www.smud.org/hp/index.html (accessed September 2001).

53. Dr. Robert Fountain, Real Estate and Land Use Institute, California State University at Sacramento, "Economic Impact of SMUD Energy Efficiency Programs," March 29, 2000, http://www.smud.org/hp/index.html.

54. The Electricity Daily, "Despite the Fuss, Some Things Work in Calif.," August 24, 2000.

55. SMUD, http://www.smud.org/hp/index.html (accessed September 2001).

56. SMUD, "Facts and Figures," http://www.smud.org/about/facts/index.html. The actual decrease was from 2,759 MW to 2,688 MW.

57. San Francisco Business Wire, "First Federal Facility Switches to 100% Renewable Power," July 23, 1999.

58. SMUD, http://www.smud.org/hp/index.html (accessed September 2001).

Chapter 9
REVERSE ENGINEERING: SOFT ENERGY PATHS

1. Robert Stobaugh and Daniel Yergin, eds., *Energy Future: Report of the Energy Project at the Harvard Business School* (New York: Vintage Books, 1983), 4.
2. We will sometimes use the term "energy conservation" simply because people use it and understand it, even though it is not precise in terms of the principles of physics.
3. Amory B. Lovins, *Soft Energy Paths: Toward a Durable Peace* (New York: Harper & Row, and Toronto: Fitzhenry & Whiteside, 1979), 38–39.
4. David B. Brooks, John B. Robinson, and Ralph D. Torrie, *2025: Soft Energy Futures for Canada*, vol. 1, *National Report of Friends of the Earth Canada to the Department of Energy, Mines and Resources and Environment Canada* (Ottawa: 1983), italics added. Note to the reader: Robert Bott, David Brooks, and John Robinson, *Life After Oil: A Renewable Energy Policy for Canada* (Edmonton: Hurtig Publishers, 1983) is the most accessible of the several versions of the Canadian soft energy path study.
5. David B. Brooks, *Another Path Not Taken: A Methodological Exploration of Water Soft Paths for Canada and Elsewhere, Report to Environment Canada* (Ottawa: Friends of the Earth Canada, 2003).
6. Brooks et al., *2025: Soft Energy Futures for Canada.*
7. Stobaugh and Yergin, *Energy Future.* The original study was reported in Demand and Conservation Panel of the Committee on Nuclear and Alternative Energy Systems, "U.S. Energy Demand: Some Low Energy Futures," *Science*, April 14, 1978.
8. See, for example, Brooks et al., *Another Path Not Taken.*
9. Ralph Torrie and David Brooks, with Ed Burt, Mario Espejo, Luc Gagnon, and Susan Holtz, *2025: Soft Energy Futures for Canada — 1988 Update* (Ottawa: The Canadian Environmental Network, 1988).
10. $30 per barrel in 2000 and $55 per barrel in 2025.
11. Janet Shawin, "Charting a New Energy Future," *State of the World: 2003* (Washington, D.C.: The Worldwatch Institute, 2003).
12. Ralph Torrie, Richard Parfett, and Paul Steenhof, *Kyoto and Beyond: The Low Emissions Path to Innovation and Efficiency* (Vancouver: The David Suzuki Foundation, and Ottawa: Climate Action Network Canada, 2002).
13. E. U. von Weizsäcker, A. B. Lovins, and L. H. Lovins, *Factor Four: Doubling Wealth, Halving Resource Use* (London: Earthscan, 1997). See also Paul Hawken, Amory Lovins, and Hunter Lovins, *Natural Capitalism: Creating the Next Industrial Revolution* (New York: Little, Brown and Company, 1999).
14. Ralph Torrie, personal communication, 2003.

Chapter 10
THE NEW BOTTOM LINE: ENERGY AND CORPORATE INGENUITY

1. For a fuller history of these tensions see D. Wheeler and M. Sillanpää, *The Stakeholder Corporation* (London: Pitman, 1997).

2. See, for example, P. Hawken, A. Lovins, and L. H. Lovins, *Natural Capitalism: Creating the Next Industrial Revolution* (New York: Little, Brown and Company, 1999).

3. J. Elkington, *Cannibals with Forks: The Triple Bottom Line of 21st Century Business* (Gabriola Island, B.C.: New Society, 1998).

4. For notable examples of contributions from senior business leaders, consider S. Schmidheiny and the Business Council on Sustainable Development, *Changing Course*, (Cambridge, MA: MIT Press, 1992); P. Hawken, *The Ecology of Commerce: A Declaration of Sustainability* (New York: Harper Collins,1993); and C. Holliday, S. Schmidheiny, and P. Watts, *Walking the Talk: The Business Case for Sustainable Development* (Sheffield, U.K.: Greenleaf Publishing, 2002).

5. Environics International, Toronto.

6. See D. V. J. Bell, *The Role of Government in Advancing Corporate Sustainability* (Toronto: Sustainable Enterprise Academy, 2002).

7. For a recent example, see C. Holliday et al., *Walking the Talk*, ibid.

8. D. Wheeler, H. Fabig, and R. Boele, "Paradoxes and Dilemmas for Stakeholder Responsive Firms in the Extractive Sector: Lessons from the Case of Shell and the Ogoni," *Journal of Business Ethics* 39, no. 3 (2002): 297–318.

9. K. McKague, D. van der Veldt, and D. Wheeler, *Growing a Sustainable Energy Company: Suncor's Venture into Alternative and Renewable Energy* (Toronto: Schulich School of Business, York University, 2001).

10. P. Drucker, "The New Meaning of Corporate Social Responsibility, *California Management Review* 26 (1982): 53–63.

11. World Resources Institute, United Nations Environment Program, and World Business Council for Sustainable Development, *Tomorrow's Markets: Global Trends and Their Implications for Business* (Washington, D.C.: WRI, 2002).

12. M. E. Porter and C. van der Linde, "Green and Competitive," *Harvard Business Review* 73 (1995): 120–134.

13. *Scotsman on Sunday*, "Heretic in the Ranks," June 9, 2002. See also H. Mintzberg, R. Simons, and K. Basu, *Beyond Selfishness* (2002).

14. See, for example, D. Wheeler, B. Colbert, and R. E. Freeman, "Focusing on Value: Reconciling Corporate Social Responsibility, Sustainability and a Stakeholder Approach in a Network World," *Journal of General Management* 28, no. 3 (2003): 1–28; and D. Wheeler, "The Successful Navigation of Uncertainty: Sustainability and the Organisation," in R. Burke and C. Cooper, eds., *Leading in Turbulent Times* (Oxford: Blackwell, 2004).

15. See, for example, C. A. O'Reilly and J. Pfeffer, *Hidden Value: How Great Companies Achieve Extraordinary Results with Ordinary People* (Boston: Harvard Business School Press, 2000).

16. D. Wheeler et al., "Paradoxes and Dilemmas for Stakeholder Responsive Firms in the Extractive Sector," ibid.

17. See J. D. Margolis and J. P. Walsh, *People and Profits? The Search for a Link Between a Company's Social and Financial Performance* (Mahwah, NJ: Erlbaum, 2001).

18. Among the best known internationally is the Dow Jones Group Sustainability Index; in Canada Innovest, BMO Jones Heward, and Michael Jantzi all produce similar indices based on their assessments of "sustainable" companies. BMO Jones Heward posted a 14 percent premium on performance for its sustainable development portfolio between August 2001 and August 2002 compared with the TSX composite index. From its inception on January 1, 2000, through to June 30, 2003, the Jantzi Social Index lost 9.75 percent, while the S&P/TSX 60 (formerly the S&P/TSE 60) lost 16.26 percent and the S&P/TSX Composite (formerly the TSE 300) lost 12.73 percent — a premium of approximately 3 to 6.5 percent for the Jantzi index.

19. See, for example, Shell International Corporate Strategy Board Research, http://www.shell.com, 43–47.

20. See the account of this process in A. de Geus, *The Living Company: Growth, Learning and Longevity in Business* (London: Nicholas Brealey, 1997).

21. Shell International, *Exploring the Future: Energy Needs, Choices and Possibilities — Scenarios to 2050* (2001), http://www.shell.com.

22. This is not a value judgement (i.e., a desired or preferred outcome in Shell's scenarios); it is simply a pragmatic assessment of what is likely based on the fact that maintaining carbon dioxide at today's levels (350 ppm) would require draconian economic and political interventions that are unlikely to be palatable to either developed or developing countries.

23. It is also worth noting here that even the Kyoto Protocol does not envisage an early return to today's carbon dioxide levels.

24. K. A. Baumart et al., *Building on the Kyoto Protocol: Options for Protecting the Climate* (Washington: World Resources Institute, 2002).

25. See, for example, United States Energy Association, "Towards a National Energy Strategy (2001), http://www.usea.org/nesreport.htm.

26. For a good review of U.S. partnership initiatives involving business, see J. N. Swisher, *The New Business Climate: A Guide to Lower Carbon Emissions and Better Business Performance* (Snowmass, CO: Rocky Mountain Institute, 2000).

27. See, for example, Government of Canada, *Climate Change Plan for Canada* (Ottawa: Government of Canada, 2002). Also available on-line at http://www.climatechange.gc.ca. See also Legislative Assembly of Ontario, Select Committee on Alternative Fuel Sources, *Final Report* (Toronto: Province of Ontario, 2002); and B.C. Climate Change Economics Impact Panel *Report to the Ministers of Water, Land and Air Protection, and Energy and Mines* (Victoria: Province of British Columbia, 2003).

28. Some commentators, such as the David Suzuki Foundation, the Climate Action Network, and Ralph Torrie, believe that an even more aggressive target could be reached by 2030, involving an approximate halving of total Canadian emissions from 727 million tonnes in 2004 to 368 million tonnes in 2030; see Kyoto and Beyond: The Low Emission Path to Innovation and Efficiency. Available from http://www.climatenetwork.org.

29. See, for example, GERT Technical Committee, "Greenhouse Gas Emission Reduction Trading Pilot Final Report" (August 2002). See also Voluntary Challenge Registry Inc., *Annual Report 2002* (Ottawa: VCR Inc., 2002).

30. See, for example, Government of Canada, *Canada's First National Climate Change Business Plan* (Ottawa: Government of Canada, 2000).
31. J. Makower, R. Pernick, and C. Wilder, "Clean-Energy Trends 2003," http://www.cleanedge.com.
32. The Cleantech Venture Network (http://www.cleantechventure.com) notes that while total amounts invested remained relatively steady in North America, energy-related investments actually dropped as a percentage of all clean technology investments, in North America from 61 percent ($103 million) to 28 percent ($89.2 million), between Q1 and Q3 2002.
33. This term was coined by MIT professor Clay Christensen to describe the problem large incumbent players have in developing technologies that will undermine sales of their existing products and services.
34. The Sustainable Enterprise Academy uses this approach very successfully in educating senior executives on business and sustainability in North America. It is described in S. L. Hart and M. B. Milstein, "Creating Sustainable Value," *Academy of Management Executive* 17 no. 2 (2003): 1–12.
35. Erroll B. Davis Jr., address to the Alliant Energy/NRDC conference on *Energy Policy and Global Climate Change – A Path Forwards*, Madison, Wisconsin, April 15, 2003.
36. A number of these companies were also award winners in the Canadian Industry Program for Energy Conservation awards presented on March 25, 2003, sponsored by Natural Resources Canada; see http://oee.nrcan.gc.ca/cipec.
37. Canadian Industry Program for Energy Conservation, *Success Stories* (Ottawa: NRCan, 2002).
38. A longer list of energy efficiency stories and benchmarks may be found at http://oee.nrcan.gc.ca/cipec/ieep/cipec/achievements/highlights.cfm.
39. Schemes such as VCR in Canada and the California Climate Action Registry in the United Sates are vital components of what may in due course become an effective system for trading carbon nationally and internationally. For without the inventories, audits, and reports of greenhouse gas emissions it is not possible to run a marketplace for carbon. We return to carbon trading in a later section.
40. Voluntary Challenge and Registry Inc., *Annual Report 2002*.
41. P. Chantraine, personal communication, 2003.
42. M. Margolick and D. Russell, *Corporate Greenhouse Gas Reduction Targets* (Arlington, VA: Pew Center on Global Climate Change, 2001).
43. Cited in World Resources Institute et al., *Tomorrow's Markets*.
44. It is estimated by the federal government that an "emissions gap" of 240 million tonnes of CO_2 will need to be bridged for Canada to comply with its Kyoto commitments by 2010 (going from a business-as-usual 805-million-tonne projected level of emissions to 565 million tonnes per annum). It is assumed that large industrial emitters will bridge just under one-quarter of the gap at 55 million tonnes. As we described earlier, some commentators envisage and advise reductions well beyond this in subsequent years, partly based on a less liberal assumption on acceptable levels of atmospheric CO_2.

45. Demand-side management (DSM) involves reducing consumer energy demands by changing behaviour or increasing the efficiency of equipment and/or processes. BC Hydro and Manitoba Hydro have a particularly successful DSM program called "Power Smart," available for industrial, commercial, and residential customers. It allows them to delay investing capital in new plants as well as export electricity to more profitable markets in the Unites States. Enbridge also has an interesting DSM program, where savings realized by customers are shared between investors and customers to enable a "win-win" scenario. Enbridge claims 2.5 million tonnes of CO_2 savings since the mid-1990s.

46. Government of Canada, *Climate Change Plan for Canada*.

47. The federal and some provincial governments already provide tax incentives to businesses for investments in energy efficiency and renewables. See, for example, Government of Canada, *Tax Incentives for Business Investments in Energy Conservation and Renewable Energy* (Ottawa: Government of Canada, 1998).

48. The *Climate Change Plan for Canada* notes "actions underway" for the transportation sector (for example, vehicle fuel efficiency improvements, ethanol enrichment of fuels, and urban transportation initiatives) that would deliver 9 million tonnes of CO_2 emission reductions. Proposed new initiatives (for example, in public transit, bio-fuels and goods transport) would deliver a further 12 million tonnes. Energy efficiency actions underway in the domestic housing and commercial buildings sectors are estimated to be capable of generating 6.7 million tonnes of savings and future initiatives a further 4 million tonnes.

49. In the United States, it is estimated that the costs of controlling acid rain through an emissions trading scheme for sulphur emissions were ten times less than originally feared.

50. The Netherlands Ministry of Spatial Planning, Housing and the Environment (VROM), http://www.vrom.nl/international.

51. R. Rosenzweig et al., *The Emerging International Greenhouse Gas Market* (Arlington, VA: Pew Center on Climate Change, 2002).

52. On January 9, 2003, the *Globe and Mail* published a story by Roma Luciw entitled "Suncor Sees No 'Material Impact' from Kyoto," which confirmed that Suncor was proceeding with all capital projects based on calculations that the impact on costs of Canada's implementation of the Kyoto Protocol would be negligible.

53. According to the Investor Responsibility Research Centre, twenty-six global warming resolutions have been filed for consideration during the 2003 proxy season, up from twenty-one in 2002. See http://www.shareholderaction.org/news/showstory.cfm?id=138.

54. D. Austin and A. Sauer, *Changing Oil: Emerging Environmental Risks and Shareholder Value in the Oil and Gas Industry* (Washington: World Resources Institute, 2002). Also available on-line at http://pubs.wri.org/pubs_pdf.cfm?PubID=3719.

55. Coalition for Environmentally Responsible Economies, *Value at Risk: Climate Change and the Future of Governance* (Boston, MA: CERES, 2002).

56. See S. O. Andersen and D. Zaelke, *Industry Genius: Inventions and People Protecting the Climate and Fragile Ozone Layer* (Sheffield, U.K.: Greenleaf, 2003). Interestingly, this book was fully endorsed by Jonathon Lash of the World

Resources Institute and Jacqueline Aloisi de Larderel of the United Nations Environment Program, and is typical of the "techno-optimist" genre.

57. See http://ca.dupont.com/dupontglobal/ca/documents/CA/en_US/pdfs/01SustGrENG.pdf.

58. See Dupont Fuel Cells, "Our Intent," *Dupont*, http://www.dupont.com/fuelcells/intent.html.

59. In the longer term, hydrogen fuel cells may replace fossil fuel engines completely; they are currently being tested in buses, delivery trucks, and personal vehicles throughout the world and will be discussed in more detail in the next section.

60. See GreenBiz.com, "FedEx Express Introduces Hybrid Electric Truck," May 21, 2003, http://www.greenbiz.com/news/news_third.cfm?NewsID=24770. Also: Clean Edge (http://www.cleanedge.com): "Hybrid electric vehicle sales are expected to exceed 500,000 units annually by 2008 and 872,000 units by 2013, according to the J.D. Power and Associates 2003 Hybrid Vehicle Outlook report. . . . Hybrid vehicles are expected to penetrate 1 percent of the market by 2005, and reach 3 percent market share by 2009."

61. DuPont and BP both have major stakes in thin-film technologies that might provide revolutionary breakthroughs for solar energy applications. See, for example, P. Fairley, "Solar on the Cheap," *Technology Review* (February 2002): 48–53.

62. Cited by the Carbon Disclosure Project. See http://www.cdproject.net.

63. See PV Manufacturing R&D Project, "Cost/Capacity Analysis for PV Manufacturing R&D Participants," *National Renewable Energy Laboratory*, http://www.nrel.gov/pvmat/pvmatcost.html.

64. See Cost of Solar Power, "Cost of Solar Power for Homes," *BP Solar*, http://www.bpsolar.com.

65. See John Kristofferson, "GE's Move into Wind Power Business Seen as Significant," Associated Press, May 16, 2003.

66. Companies like Westport Innovations (B.C.), which makes high-efficiency fuel systems, and EnWave (Ontario), which uses deepwater circulation systems for heating and cooling buildings, have successfully developed technologies that demonstrate Canadian capabilities for taking novel energy products to market. See also the quarterly reports of the Cleantech Venture Network (http://www.cleantechventure.com) for recent venture capital deals involving Canadian energy technology companies.

67. Legislative Assembly of Ontario, Select Committee on Alternative Fuel Sources, *Final Report*.

68. See http://www.pewclimate.org/belc/shell.cfm.

69. See, for example, BP, *Lower Carbon Growth Story: Reflections on the Interaction of BP with the Environment — Acting on Our Convictions* (London: BP plc, 2002).

70. According to the 2002 International Energy Agency (IEA) report, *Renewables in Global Energy Supply*, renewables currently provide just 13.8 percent of world total primary energy supplies.

71. The World Resources Institute has produced a range of very helpful *Corporate Guides to Green Power Markets*, all available from http://www.thegreenpowergroup.org.

72. See article by U.K. Minister of State for Foreign Affairs Peter Hain in *Greenpeace Business* 59, 2.
73. See, for example, M. T. Jones, "Towards a Hydrogen Economy," (presentation to the IEA Renewable Energy Working Party Seminar, Paris, March 2003); or C. Van Zeyl and T. Kimmel, "Facing Up to the Challenge: Hydrogen Infrastructure for the US and Canada" (presentation to the EECO Conference, Toronto, May 2003).
74. For an example of upbeat, urgent U.S. commentary predicated on technological and security assumptions see P. Schwartz and D. Randall, "How Hydrogen Can Save America," *Wired* Supplement, April 2003, 1–13. For a more sober technical and economic analysis see J. N. Swisher, *Cleaner Energy, Greener Profits: Fuel Cells as Cost-Effective Distributed Energy Resources* (Snowmass, CO: Rocky Mountain Institute, 2002).
75. Many of the people we interviewed noted the barriers generally to renewables becoming viable, including the absence of energy storage technologies and infrastructure (IT, refueling, connections to grid, etc.).
76. See R. Mang, "Canada Makes Its Mark on the Hydrogen Planet," *Globe and Mail* Special Information Supplement, June 9, 2003, H4.
77. J. Edmonds, M. J. Scott, J. M. Roop, and C. N. MacCracken, *International Emissions Trading and Global Climate Change: Impacts on the Costs of Greenhouse Gas Mitigation* (Arlington, VA: Pew Center on Global Climate Change, 1999).
78. R. Rosenzweig, M. Varilek, and J. Janssen, *The Emerging International Greenhouse Gas Market* (Arlington, VA: Pew Center on Global Climate Change, 2002).
79. TransAlta's and OPG's experiences in carbon trading, and indeed those of the U.K. and elsewhere, were explored in more detail at a conference called "Market Mechanisms for Emissions Trading and Reduction," held in March–April 2003 in Toronto at the Canadian Institute.

The authors would like to thank all the interviewees and providers of primary information — they were true representatives of the spirit of a new age. We especially would like to acknowledge the thoughtful and extensive contributions of Mark Weintraub of Shell International, John Mogford of BP, Norm Lockington of Dofasco, Gord Lambert of Suncor Energy, Peter Chantrain of DuPont Canada, Bruce Lourie of the Richard Ivey Foundation, and Brian Kelly of the Sustainable Enterprise Academy.

Chapter 11
LIFE AFTER OIL: A NIGERIAN EXPERIMENT

1. Ogoni Bill of Rights, 1990,
   http://www.pbs.org/newamericans/2.0/nigeria/sarowiwa2.htm.
2. "Nigeria To Use New US Ship To Curb Crude Oil Smuggling," Petroleumworld.com, September 5, 2003,
   http://www.petroleumworld.com/story2259.htm.

Chapter 12
IDEAS FOR THE HOME FRONT

1. Office of Energy Efficiency, *Business Plan 2001–2002* (Ottawa: Natural Resources Canada, 2001).
2. Avi Friedman, "The Grow Home: Design Strategies for Low Energy Consumption — A Case Study," *Energy & Environment: Proceedings of the 17th Annual Conference of the Solar Energy Society of Canada,* Toronto, Ontario (June 21–26, 1991): 93–98.
3. Canada Mortgage and Housing Corporation (CMHC), *Tap the Sun: Passive Solar Techniques and Home Designs* (Ottawa: CMHC, 1997).
4. Godfrey Boyle, *Renewable Energy: Power for a Sustainable Future* (Oxford: Oxford University Press, 1996).
5. Mario Chiarelli, *Energy-Efficient Housing: Reference Guide* (Toronto: Ontario Hydro, 1989).
6. See note 4 above.
7. Ibid.
8. Canada Mortgage and Housing Corporation (CMHC), *Energy Conservation in New Small Residential Buildings* (Ottawa: CMHC, 1981).
9. Witold Rybczynski, Avi Friedman, and Susan Ross, *The Grow Home* (Montreal: McGill University School of Architecture Affordable Homes Program, 1990).
10. ASHRAE, *ASHRAE Handbook: 1989 Fundamentals* (Atlanta: ASHRAE, 1989): chaps. 23, 27.

Further Reading: Avi Friedman, *Innovation and the North American Homebuilding Industry* (Montreal: McGill University School of Architecture Affordable Homes Program, 1990).

Chapter 13
GENERATION 2023: TALKING WITH THE FUTURE

1. The cousins mentioned in this piece are some of my real grandchildren, though in twenty years they may be surprised at what I once thought they'd be doing and at how likely I thought they'd be to ask questions which by then would have been common. Everyone else mentioned is also real. I am indebted to all my colleagues in the venture business, academia, and fuel-cell companies, and especially to the management team and Industry Advisory Committee at Chrysalix, for the ideas that have gone into this paper. However, I accept complete responsibility for all of its opinionated content!
2. Christine Bergeron, Investment Manager, Chrysalix Energy Management Inc., prepared the Chrysalix Roadmap.

Afterword

1. Garrett Hardin, "The Tragedy of the Commons," *Science* 162: 1243–48.

EDITORS' ACKNOWLEDGEMENTS

This book began as a question: is it possible to find a sustainable way to fuel the future? The people we found to answer this question have our deepest thanks for their extraordinary effort, dedication, and ingenuity: Dr. Geoffrey Ballard, Len Bolger, Dr. David B. Brooks, Michael J. Brown, Peter Fairley, Dr. Avi Friedman, Susan Holtz, Thomas Homer-Dixon, Dr. Eddy Isaacs, Wyatt King, Gordon Laird, L. Hunter Lovins, Allison Macfarlane, Jeremy Rifkin, Jane Thomson, Dr. David Wheeler, and Ken Wiwa.

We should single out Thomas Homer-Dixon in this group, as his work inspired this project. He is truly one of Canada's priceless resources.

And for his research and work on the history of energy timeline, Ian Connacher deserves gratitude. And Cathy Gulli's fact-checking of crucial sections was vital to the book's accuracy, though any errors are ours, not hers.

A collection of essays comes together because of a wide collection of people, and a few merit special attention. Jan Simpson and Don Simpson introduced us to so many people and ideas from Alberta and did much to build trust between diverse groups of people who too often let their differing visions of the future stop the much-needed sharing of ideas.

Rob Steiner, Tammy Quinn, Barry Gordon, John Ketchum, and Carol Divine were all part of the original meeting at C'est What and their ideas seeded the ground for this project.

Martha Sharpe grasped the idea for this book and for the Ingenuity Project series immediately and gave us our first and our most constant encouragement to complete it. Dedicated, smart, and insightful, she is what a publisher ought to be. Kevin Linder's thoughtful editing and probing questions made this collection into a complete book. And Laura Repas deserves credit in helping to promote and publicize the book.

As well, the *Ingenuity Project: Fueling the Future* involves many people, including Bernie Lucht of CBC Radio's *Ideas*, Jennifer McGuire, and Neil Sandell. From CBC TV, Tony Burman, Maria Maronwiz, Heaton Dyer, Julie Bristow, and Stuart Coxe have all been crucial, Mike Downie has brought his great talents to the television documentary, and Ann Elisabeth Samson has been invaluable in helping to develop and organize the television project. At CBC Online, Mark Hyland, our great friend and long-time business partner, has, as always, offered us his sage advice, and even more importantly his energy and time. And Claude Galipeau and Fred Mattocks from CBC Online have been instrumental as well. From *Maclean's* magazine, Michael Anthony Wilson Smith and Michael Benedict were instrumental and helpful in providing a broader venue for these ideas.

And as always, our friend, advisor, and legal council Michael Levine has been at the centre of this project, as he has in so many of our — and so many other's — endeavours.

Finally, and most importantly, thanks to our wives, Tammy Quinn and Rosalind Heintzman, for your ideas, love, and generosity, without which none of this would happen.